The Knowledge Game

The Revolution in Learning and Communication in the Workplace

edited by
Gordon Wills

Cassell

Wellington House
125 Strand
London WC2R 0BB

First published in 1998

British Library Cataloguing-in-Publication Data
A catalogue record for this book is available from the British Library.

ISBN   0-304-70325-7 (hardback)
          0-304-70326-5 (paperback)

Typeset by MCB University Press, Bradford, West Yorkshire
Printed and bound in Great Britain by Redwood Books Ltd, Trowbridge, Wiltshire

The Knowledge Game

The Revolution in Learning and Communication in the Workplace

edited by
Gordon Wills

CONTENTS

Abstracts and keywords _____ 5

Preface _____ 9

1. **E-postcards from the other side**
   *impressions sent home to Trad University while on
   sabbatical leave in Internetica* _____ 11

2. **Embracing electronic publishing**
   *argues there is no point in resisting the new paradigm:
   embrace it* _____ 44

3. **Re-engineering knowledge logistics**
   *logistics flow modelling explores things to come
   – with Mathew Wills* _____ 64

4. **The ins and the outs of electronic
   publishing**
   *explores the new phenomenon and how it's happening
   – with Mathew Wills* _____ 80

5. **Author authority in the ascendant**
   *argues that authors have new roles to play
   – with Chris Wiles* _____ 96

6. **Rogue learning on the company
   reservation**
   *Tom Reeves compares action learning at work in
   Seagram and MCB University Press* _____ 108

7. **Leading courageous managers on**
   *explores in depth the action learning processes
   and outcomes at MCB University Press
   – with Lesley Gore and Kathryn Toledano* _____ 122

8. **ROI in management development**
   *evaluates the benefits of action learning around the globe
   – with Carol Oliver* _____ 136

CONTENTS
(Continued)

9. **Networking and its leadership processes**
   argues that networks are not easy to lead and that little is known _____ 144

10. **Designing a quality action learning process for managers**
    shows ISO 9002 readily delivers democratic goals
    – with Molly Ainslie _____ 158

11. **Creating a marketing intelligentsia**
    identifies the origins of database marketing
    – with Bev Bruce and Timmie Duncan _____ 172

12. **Journey to marketing clubland**
    originates the notion of communities of interest
    – with Julian Wills _____ 199

13. **Learning in marketing clubland**
    captures learning within communities of interest
    – with Julian Wills _____ 223

14. **Realizing the benefits of a marketing intelligentsia**
    evaluates how far database marketing has come on the threshold of electronic publishing
    – with Bev Bruce and Tracy Jordan _____ 244

1. E-postcards from the other side

Gordon Wills

Keywords Business schools,
Communities of interest, Distance learning,
Electronic publishing, Quality,
Virtual conferencing

Records the substantial leave experiences of George, Professor of Marketing, and Margaret, Quality Consultant, in Internetica where electronic publishing and the Information Society are now coolly embraced. Their e-postcards home demonstrate that some of our current scepticism is well founded as illustrated by lower than hoped for active participation rates and the challenge of transitory structures as contributors to learning. But they predominantly capture an emerging maturity among all the excitement that is creating the requisite discipline to focus on what can be enduringly achieved over meaningful time-scales. Both George and Margaret make up their minds to work on a few specific high value added applications areas on their return home – most particularly the intelligent capture and deployment of high quality information and the design and sustenance of user communities of interest.

2. Embracing electronic publishing

Gordon Wills

Keywords Electronic publishing,
Organizational change, Technological change

Describes how MCB University Press has started to come to terms with the metamorphosis which electronic publishing offers. Sees a future where interactive multimedia products and services are the norm and are quite differently distributed, based on new alliances from within but also from outside the traditional players. Explores how MCB's strengths might be used to succeed in the new frameworks and concludes that double-loop action learning is the only viable way ahead. Identifies that authors will be a constant point of reference and that networked desktop PCs and networked homes will open vast new markets to those who can re-present knowledge and information to gain and hold their attention. Outlines what MCB has done so far.

3. Re-engineering knowledge logistics

Gordon Wills and Mathew Wills

Keywords BPR, Knowledge workers,
Logistics

Describes the key elements of total logistics systems and their cycle times for requisite service levels at least cost. Shows how these constructs originally emerged from military necessity but have more recently been driven for commercial and manufacturing advantage. Analyses the traditional logistics cycle in academic and professional publishing and then demonstrates how the application of a total logistics system approach with the emerging capabilities of electronics totally transforms the performance of the system, reducing cycle time by 75 per cent. Significantly re-engineers the five key elements of logistics systems – facilities, unitization, communications, inventory and transportation – and rewrites the cost/benefit equation of service levels. Explores the opportunities for backward and forward integration by traditional librarians and publishers respectively in the re-engineered total system.

4. The ins and the outs of electronic publishing

Mathew Wills and Gordon Wills

Keywords Customers, Electronic publishing,
Internet, Marketing strategy

Electronic publishing needs a strong input of marketing thinking. Technological hype has created a sales fetish which has little evidence to support its claims. The substantive benefits when a broader perspective is taken for authors and readers are very significant, including considerably faster publication and much wider dissemination via Internet. Archival knowledge and current awareness/

browsing of the body of knowledge and information require quite different marketing approaches. Little attention has been given to their discrete needs. Draws comparisons from retailing theory and from the emerging range of experimental cases from Internet pioneers to identify robust strategies for short- and medium-term action by publishers. They imply a determined effort to avoid hard selling and product-driven mindsets in favour of exploitation of the scope for interactive and integrated marketing to authors and readers alike.

5. Author authority in the ascendant

Gordon Wills and Chris Wiles

Keywords Corporate strategy,
Customer requirements, Journal publishing

The buyer's market in traditional academic and professional publishing is clearly at great risk from the arriving electronic capabilities. Publishers wishing to survive and prosper in the new paradigm must cultivate deep understanding of authors' needs and of their own abilities to add value amid the new technologies, rather than being mesmerized by them. A leading learning strategy has been followed by MCB University Press since it established the Literati Club for its authors and editors in 1991. What began as a search for greater production and marketing efficiencies rapidly evolved into a vehicle for effective productive reasoning about the corporate future. Five years later it has positioned the company as the most comprehensive world-class publisher on the Internet. The sequence of leading actions and double loop learning is described and analysed.

6. Rogue learning on the company reservation

Tom Reeves

Keywords Action learning,
Corporate culture, Management development

Compares two companies' use of action learning, one primarily for individual staff development, the other for staff development and business objectives. In the second, action learning's questioning, problem-solving ethos had pervaded corporate life. This innovative culture was substantially driven by top management, but this was not a free licence for staff to innovate, and learners' initiatives were restrained within implicit boundaries. The research, covering all levels, was conducted by interviews and group discussions. Recommendations to amplify individual learning for corporate impact include: develop staff to bring about change; cultivate a receptive culture; allow time; tolerate some subversiveness; and prevent bounds to autonomy from inhibiting initiative.

7. Leading courageous managers on

Lesley Gore, Kathryn Toledano and Gordon Wills

Keywords Action Learning,
Electronic Publishing, Strategy

In the light of impending changes that will affect the MCB publishing enterprise, explores the leadership strategies which the owners have pursued. These comprise: management action learning; systems development; mentoring and coaching, and structured change.

8. ROI in management development

Gordon Wills and Carol Oliver

Keywords Action learning, Business schools,
Management development,
Masters of Business Administration,
Return on investment

Too few programmes of management development seek to evaluate their hard ROI for the enterprise. Describes how action learning's focus on company-specific issues makes this more feasible. Reports on a four-year impact analysis from MBA programmes showing that employing organizations benefited greatly and that the individual managers also gained a host of soft benefits. The endemic problems of action learning are also identified but the contribution of the Set (fellow members of a small learning cell) again stands out as most vital.

9. Networking and its leadership processes

Gordon Wills

Keywords Empowerment, Learning, Management, Networking

Aims to review what has been learned and published recently about networking, using examples from the experiences of the International Management Centre (IMC). Describes the IMC network and leadership process (including the use of the two-level Bulletin Board System and the Atherton Intelligence System) before discussing why networks are necessary and detailing how they function. Covers areas such as finance, network evolution, and the major causes of failure. Concludes that networks are an important tool for the future in view of their ability to change to meet market needs.

10. Designing a quality action learning process for managers

Molly Ainslie and Gordon Wills

Keywords Action learning, Internet, Managers, Quality assurance

Describes the evolution of an Internet-driven dynamic quality assurance system for action learning programmes across the world. It replaced a relatively inefficient paper-based process and was preceded initially by Bulletin Board/electronic data interchange procedures. Its success was imperative for a global Business School to comprehend how Sets were proceeding while avoiding "controlling" processes that contradict the action learning paradigm. This was further reinforced in the joint venture between International Management Centres and the University of Surrey, as both were required to meet quality assurance monitoring requirements of disparate agencies globally. When the procedures gained ISO 9002-accredited status, there was an upsurge in concern to improve further the patterns of faculty induction and continuous training and development for their facilitation skills through faculty development scholarships and delivery

effectiveness workshops. The approach is now operative for North America, Africa, Asia Pacific, Australasia and Europe.

11. Creating a marketing intelligentsia

Gordon Wills, Bev Bruce and Timmie Duncan

Keywords Customer loyalty, Database management, Marketing strategy

Database marketing is scarcely comprehended by marketing professional practitioners as yet but its potential is becoming clear. It not only revolutionises the productivity of direct marketing but transforms the effectiveness of salesforce prospecting, marketing research and channel management. The evolution of database marketing at MCB University Press is described and analysed over the past 15 years. The intrapreneurial and behavioural dimensions of development and implementation are traced and guidance provided for those who accept that good computer systems grow from within rather than arrive from without the enterprise.

12. Journey to marketing clubland

Gordon Wills and Julian Wills

Keywords Customer loyalty, Database management, Marketing strategy

The emergence of marketing clubs has been made possible by database technology. Reviews the published literature in depth and categorizes clubs from suspects to regular subscribers. Analyses three new clubs launched by MCB University Press for readers and authors showing achievements to date.

13. Learning in marketing clubland

Gordon Wills and Julian Wills

Keywords Data collection, Database management, Human resource management, Librarians

Compares the launch and development, by an academic publishing house, of three marketing clubs, two of which were focused on customers and the third on authors and editors. Promotional campaigns achieved good results but, most importantly, the marketing intelligence gained transformed the organization and prime marketing structures. Fresh objectives for future growth were identified and computer systems further developed.

14. Realizing the benefits of a marketing intelligentsia

Bev Bruce, Tracy Jordan and Gordon Wills

Keywords Customers, Database marketing, Electronic publishing, Organizational development, BPR

MCB University Press has conducted data-based marketing campaigns since 1972, but has escalated its investment as new technologies and software have become available. In-house staff have now designed and introduced an integrated customer environment (ICE). That affords not only major productivity gains but also a better platform for creative marketing thinking. Customer clubs have evolved for librarians and human resource professional managers and for authors on the supply side. Intelligence has been derived to direct worldwide field visit programmes to key customers and authors. And the relationship developed with customers provides the foundation for the company's emerging strategies for customer-led product design in electronic publishing.

Preface

In his Foreword to my last collection of essays published as *Marketing Reality,* (1991) MCB University Press, John Peters described them as "personal and eclectic, exploring the application of marketing thought to areas where it isn't always applied". That was indeed my intention then and it is continued further here.

I have not extrapolated my learning as so many of my erstwhile marketing colleagues have done since I first became a Professor of Marketing at the University of Bradford in 1968. Accordingly I have not contrived a need to be post-modern. I still find the earliest works of Wroe Alderson inspirational and I still find the 4Ps and the product life cycle valued touchstones as Stravinsky reputedly did the chromatic scale. My learnings have led me further into the concept of customers, of demos, as increasingly powerful players in their places of learning and of working and within marketing itself.

Until the arrival of electronic publishing, with its democratization of knowledge and its concomitant universal access to the warehouse of knowledge in the library, I had not discerned the connection which I seek to advance here. But I believe I do now. I believe that empowerment and the action learning paradigm in the workplace, the evolution of database marketing that respects the rights and individuality of the customer, and electronic capture and dissemination of knowledge are all manifestations of democratic action intellectualized and legitimized in western societies by the emergence of marketing as a philosophical discipline since the 1960s. Indeed, I would go further and argue that the new workplace for most of us will be the intensely democratic physical reality of our own home and the virtual reality of electronic communities of interest.

The essays presented here come under those three broad areas of democratic action. And, as must surely be envisaged, what has been engendered was done with many other colleagues and friends. Where they have worked with me to write the words I gladly and rightly acknowledge their authorship and thank them for their partnership.

In *Marketing Reality* I cited Adam Smith's dictum: "the sole end and purpose of production is consumption". Here I would like to cite another equally perceptive comment: "sales are limited by the extent of the market". The democratization of knowledge, which Adam Smith also focused on extensively, as it explores and develops who we are at work, who we are in the marketing database and who we are in the face of knowledge itself, has seen a multitude of walls coming down. It is an event of greater significance than the tearing down of the Berlin Wall but that symbolic event has amply illustrated just how much can and indeed must change when walls come down.

But I want my final words to explain why I so greatly value action. I have truly grown weary of the chattering classes, just as they have wearied of my determination for action. I have shamelessly incited and engendered action as a wholly worthy route to discerning how customers can learn to understand and learn to enjoy. In doing this I have been incomparably encouraged and supported by the team at Buckingham led by Carol Oliver and Molly Ainslie, including most recently young members of the burgeoning Internet Research and Development Centre. To them all I dedicate this collection of essays.

Gordon Wills
Milton Malsor
July 1997

Chapter 1

E-postcards from the other side

George and Margaret had leapt at the opportunity-for-two to take a year's sabbatical leave in Internetica when they saw it advertised on the Net. They were selected from more than 8,000 applicants because their interests were both pursuable in the capital, Internorbert. George was Dean and Professor of Marketing at the Business (B) School at Trad University (U) and Margaret a Founding Partner of Relationship Consultancy Inc. They both despaired of the negative obstructionism and contrived delays of colleagues at work, who they characterized as fearful rather than wise. They resented the suggestions that what was on offer would automatically undermine traditional quality standards and values. They pined for a spell among believers in the value added benefits of the Information Society where they could talk and enjoy the electronic leverage they could see blossoming all around them.

Once they had been accepted by Internetica, they were surprised by the alacrity with which colleagues had agreed it would be good for them to get away. Only Fredrick, the immediate past Dean of Business and a family friend of them both, had wryly told George his sabbatical leave proposal had been enthusiastically agreed so that Trad U could be spared the constant reminders that something needed to be done before too long for at least one year longer. But George's colleagues had been wise enough to ensure he was committed, as was Margaret when her partners far more reluctantly agreed to her leave, to produce a Considered Report and Personal Action Plan on their return (see Faxbacks 2 and 3).

The correspondence continues after their return as fresh issues arise.

The e-postcards and faxbacks between Internetica and Trad U

E-postcard 1   All quiet here, quieter in fact than expected

E-postcard 2   Nothing's quite finished here – a bit like Taiwan really

E-postcard 3   Internet quality at last?

E-postcard 4   This looks like paradise for faculty members

E-postcard 5   Doctorates are getting more straightforward to create

E-postcard 6   A nice consultancy assignment at last …

E-postcard 7   Getting ahead of the moment of publication and myself …

E-postcard 8   Central place theory and social physics – well fancy that

FaxBack 1   You're underestimating us back here

E-postcard 9   European Union and net gain quite right

E-postcard 10 PEERNet gets closer to home for authors

These e-postcards and faxbacks continue forward @http://www. irdc.com/

Anbar Management Magazine,
No. 2, June 1997

E-postcard 11 The Dean is determined to launch an e-journal from U of I

E-postcard 12 What can we hope to transfer back for Margaret and Trad U?

FaxBack 2   The George Report to Trad U

FaxBack 3   The Margaret Report to Relationship Consultancy Inc.

E-postcard 13 I dreamt I was a global tutor

E-postcard 14 Internet R&D Centre's secret weapons

E-postcard 15 Emergent global columnists on the Net

E-postcard 16 Emergent master teachers on the Internet

E-postcard 17 The quest for business models

E-references and hyperlinks

POSTCARDS FROM
THE OTHER SIDE

E-postcard 1: All quiet here, quieter in fact than expected

"*Fredrick.* We just arrived safely. First impressions are how very quiet it is. Not a lot of activity. We arrived over the weekend of course so we presume everyone is out of town but we thought they'd all be cyberactive somehow. Like a Hong Kong or Singapore cellular phone circus in every street and every restaurant and at every meeting.

Very convenient arrangements at the airport. We had a chance to evaluate the intelligent apartment we had booked and the comparative ratings given lately to all at a similar price and promptly changed our minds. You'll be amused we decided to go for a "technology-withdrawal suite" where all we have is satellite movie access in the lounge and recipes on-line for the kitchen. The proposition that convinced us was it was nice to get away from all the technology in the evenings after being submerged all day. What hope for working from home then in the future?

Reception was impersonal at the apartment nonetheless. No friendly character for a chat about what to do and where to go. All on screen in the apartment! Had to let the robot swipe our smartcard to get in the place at all.

One fascinating thing you'd also like. The lounge is full of books – all outside our areas of interest now but by the end of the year I am sure we'll know them inside out. They're on art history in France and Holland, resistance to new ideas everywhere, and yards of English poetry and literature. There's a logic to the collection which will get through to us soon I am sure. Either it's to relax or reassure.

Must dash and consult the recipes on-line again. We are finding it hard to decide with so much choice. Can't get over the low level humm or is it the absence of real buzz? That's taken us completely by surprise. We will share it with our new colleagues next week and see what explanation they might have.

Ciao – G&M".

***E-postcard 2*: Nothing's quite finished here – a bit like Taiwan really**
"*Fredrick*. Its been a very busy first few weeks. I have settled in well at the Internet Management Centres (IMC) at the University of Internorbert (U of I), and Margaret at the International Quality Management Association (IQMA) which is resourced from here. Both operate 100 per cent on the Net.

The biggest resources everyone uses professionally in Internorbert are familiar names from Trad U such as EBSCO, Blackwells and Dawson Faxon. They have monstrous e-warehouses where access is available to all and sundry using smartcard referencing. We haven't got a library as such, just access everywhere you look – except mercifully at home where the books are proving a blessing already.

I can hardly recall the names of journals now in the marketing field. I make use of extensive databases covering the world's top titles and search on key words, quality star-ings and document type.

The big resource e-warehouses emphasize that the content they disseminate is no better than the original article published somewhere previously, but little mention is made of those original publishers. I was surprised not to see some of them here with big resource centres, but no. They must be somewhere else doing something else. Margaret thinks they are hooked on archives and databases at the expense of user behaviour patterns. She's normally got a good nose for these things and it certainly fits in with marketing theory of technology and product-driven manufacturing.

Nothing's quite finished though. Rather like our first impressions of Taiwan as we took that taxi ride from the airport to downtown Taipei. Remember? We wondered why all the buildings looked half completed and ill maintained. Here it's the electronic structures that lie unfinished/pages "not found", or long since last updated. We thought that would have been readily overcome here in a city where we have nothing but Internet believers. But the truth seems to be that those who launched the great e-enterprise are lacking the skills to keep it all shipshape. A different sort of talent is required to ensure week-in week-out quality of e-resources. Takes a different sort of talent to manage the whole thing with priorities too. Sounds familiar?

Best wishes G".

***E-postcard 3*: Internet quality at last?**
"*Fredrick*. Time I wrote, eh? George dropped me off on Day 1 at the International Quality Management Association (IQMA) who have me under their wing for the year. Amazing place. They have 25,000 members already paying $US100 per annum subscription and provide a flourishing on-line professional educational, recruitment and consultancy service. I have of course been one for a couple of years but had no appreciation of the fractal depth and breadth of their activities.

They were very kind Day 1. I thought I put my foot in it by saying how glad I was to see them face-to-face (f2f in Internetica speak) but they quite understood. Said it was common at first, common in the middle and common all

These e-postcards
and faxbacks
continue forward
@http://www.
irdc.com/

the time … Communing with PCs full time they felt was not likely to be a "total" quality experience and the tactile media needed to get their act together as complementaries rather than dinosaurs. Think they were being polite.

Anyway, they had some comments on George's and my surprise as reflected in E-PCs 1 and 2 on the quiet and the unfinished nature of much around. The quiet is truly low participative activity levels; only so many hours in a day and how do you want to spend them? But they do make the point that an awful lot of casual traffic passes by and you do not hear or see it. Sites can have only ten contributions for 1,000 visits and that can be regarded as satisfactory by all concerned. The ten had their say, and the 990 saw and heard what they wanted. Some lingered; others were in the wrong place and left swiftly.

On the unfinished structures which we cannot help noticing all the while, they are very philosophical, saying that the next and coming wave is from precisely the finishers, concluders, upholders and maintainers. Such folk never go on to the beaches first; they get ashore as soon as possible afterwards and get the place tidied up. There is, to tell the truth, a new division here in IQMA that has just started using ISO 9002 to discipline and structure our own work. Serve us right. We are awash with non-conformances as soon as we set the quality standards but most of the deviations can be put right as routines. The more creative folk who design the resources and then cut and run are surprisingly welcoming this initiative, which makes a change from the derogatory remarks normally heaped on ISO 9000 enthusiasts or even auditors back where you are Fredrick.

I've fallen on my feet here at IQMA anyway. They said if as a consultant I wanted more edge-of-the-seat buzz at a site than elegantly contrived archival access, I had better turn my mind to it quickly. Others it seems are reaching a similar conclusion. The IQMA is seeking to develop more interactive services such as virtual conferencing (Heal, 1996) in the hope of getting cost effective linkages and lateral thoughts. But they are not overly optimistic about great results because of the earlier view that a lot of attendees do not want to contribute. They just want to visit and look and maybe learn … and why not? Nothing new in that at all but we all wonder whether the Internet has got some ingredient that can set a new pattern alight. It's going to require some lateral thinking. The analogy they make here is that an important part of any traditional conference is f2f in the bar, lobby and the like, but we have not seemingly captured that moment for interaction and benefit sharing.

The notion I shall be working on is to use electronics for the before and after stages of good f2f sessions and also to capture views from those who know they cannot come f2f but by targeting or shoulder tapping as it's called over here. This scenario suggests a booming role then for f2f with electronic enhancements. Perhaps that's a worthwhile complementary media pattern, eh? The refreshing thing is how dissatisfied everyone is intellectually with the staggering progress they have made. From a quality perspective, the place is alive with continuous improvement. George's U of I has got several research

programmes running on the social-psychological aspects of the Internet itself
searching for its limits instead of its boundless opportunities!

To try and get a better understanding I have been joining a lot of Quality
conferences and discussion/listserves/forums. It seems crystal clear that the
technology has to improve its speed and that we will have to wait for greater
visual and/or aural connectivities, before they really take off. Currently,
reading all the contributions at a serious conference is unrealistic for any but
those most deeply involved. The best that can be created is a good ongoing
argument/ debate where responding to the last point made is a worthy
participation for most. George I know agrees with open ended discussions but
he has already seen at IMC in the B School at U of I what can be accomplished
by a small set of managers and faculty with a precise and deadlined focus to
work towards. There archive, f2f and virtual conferencing can all be deployed
effectively.

Hopeful now that by the end of the year I shall be smart enough to make
something sensible work back at Relationship Consultancy Inc.

As ever M".

E-postcard 4: **This looks like paradise for faculty members**

"Fredrick. Margaret tells me it's my turn to write. She's ankle deep in the ISO
9000 Division's work now at the IQMA. Seems clients are queuing up to get
structured and audited and accredited – not for their customers but for their
own sanity. That's a refreshingly new angle on ISO 9000 (Peters, 1996).

Here at the U of I the Dean turned out to be a lady. Most energetic, good
scholar too in logistics information systems. She's currently doing the ISO
9000 thing for the B School (Ainslie and Wills, 1997) and using the Charter
approach – for the students, the faculty, the graduates and whoever else the B
School might serve. Its brilliantly simple made possible by the past three
year's work putting the entire activity and all its resources on to the Internet
site. And the whole is linked to the global research library for managers
known as ANBAR Electronic Intelligence.

She was going over the six-monthly literature updating of the marketing
and logistics core programmes in my first week and asked me to join in. The
amazing thing is that by using ANBAR Electronic Intelligence the task can be
done in half-a-day at most. She described the whole process as follows and it
must be the future.

First and foremost, all programme structures are located on the Internet.
Then each and every module of study is allocated keywords in the literature
to describe that module's focus. She was at the time looking at "database
marketing" and that showed 104 abstracts in recent literature, which can of
course be sorted by a host of useful classificatory criteria ANBAR has
ascribed to each one.

Anyway, the model in use is one that could never be envisioned without
such on-line links. At IMC when a student searches the programme structure
and views the keywords associated with each module, a hit on the link gives

POSTCARDS FROM
THE OTHER SIDE

These e-postcards
and faxbacks
continue forward
@http://www.
irdc.com/

initial access to the most recent faculty selection, and then beyond that to the full archive of 104. And those abstracts cover the top 400+ journals in the world and the access is never more than one month lag behind publication.

The catch is that in most cases, 320 out of 400 journals, only the abstract and not the full text article as well, is available on-line. But the number is rising all the time (Morris, 1995). The B School, IMC, has an inclusive access subscription in its fees for all the abstracts and a modest 25 full text allowance as well, but beyond that smart/credit card payment on access is required. If they are not yet available on-line, then two-hour fax back is offered or ye olde postal service, dispatched within 24 hours both from the British Library.

The Dean here confesses she never now goes to any focused workshop/ seminar/lecture/tutorial without a quick search on several relevant keywords just to note where the most recent articles are and what they have to say in abstract form. For a consultancy visit she will normally search under the "practice implications" high scoring articles within the keywords. She reckons that she retrieves full text for one in five abstracts, making an average of 200 or so each year, which currently costs her just short of $US 3,500. She has and uses a U of I smartcard for the first $US 2,500 but after that it is normally to her own account.

The great majority of that full text cost goes to pay royalties to the primary publisher and some of them are pushing it upwards. They are scared stiff that the sales of single articles will decimate their big subscription income and surely they are right. But the consensus here, albeit sympathetic to their plight, is that it is madness to push up royalties. Authors will shortly demand a share of them and the true benefit of electronic publishing which can and must be high volume/low price dissemination will be squandered. The unanimous view is that authors should insist that the royalties are kept to an absolute minimum

and that through low full text retrieval fees the maximum number of individuals can have access. Provided the publisher can believe in the future of the Information Society, the investment payback will be there beyond doubt if not the fast buck.

I asked the Dean how they dealt with text books for students at U of I, and it appears they have three approaches. First, they did away with them and made the course materials give the structure fleshed out with the latest articles, which is fine for graduate programmes. Second, they retained them, noticeably on undergraduate programmes, and simply provided a direct link to the nearest on-line bookstore to ship/deliver to them with smartcard debiting. Third, and this is the growing sector, the U of I is serializing some of the best textbooks, chapter by chapter, with smartcard payment by each chapter downloaded. With portable document formatting (PDF) technology and colour printers it works well if still somewhat slowly. It certainly beats ye olde post office in outlying areas or corners of the world.

This whole sequence is one big eye-opener for me and shows what the future can offer for us at Trad U too. Best of all is the talk here about getting back to small sets of students/managers working f2f as often as possible and as

CyberSets too with Web Forums. Just like the good old days; tutors will be able to work with students as individuals as well as moderating effective peer learning and support sets.

Best wishes G."

E-postcard 5: Doctorates are getting more straightforward to create

"Fredrick. Margaret was fascinated to see what I wrote to you on my last card about direct keyword accessing and the imaginative search criteria. She has been promising herself to do a practitioner doctorate or at least a short state-of-the-art book for as long as I can remember. She's been off and registered here at the U of I to do one on the strength of two points really. First, tracking down the literature is now transformed from a chore to straightforward hard work. Second, the U of I offers the opportunity to make use of the wide range of her consultancy assignments and outcomes therefrom as the data together with a linking explication as they call it towards their Doctor of Business Administration (DBA). She is expected by the U of I to create a series of papers for ongoing publication during her studies and, as appropriate, a book on completion. Two birds with one stone, eh.

She's already got IQMA and some of its associated journals to convene a virtual conference on her theme which is, of course, on the application of ISO 9000 to electronic publishing and management education.

Fredrick, this is Margaret. I have taken over the pen. Cannot have George telling you all the good news on the DBA and book front. Can you imagine how delighted I was when I discovered that ANBAR Electronic Intelligence classifies all its abstracted papers by document type as well as content type. So I got cracking with 16 out of 800 recent papers being literature reviews of my area immediately to hand and with a wealth of references from earlier years that can be traced back via the British Library as well, albeit more slowly. With this type of resourcing, I can overcome my darkest fear at doctoral level that somewhere out there is an important article I have missed.

My virtual conference has already picked up contributions from New Zealand on how the US has used ISO 9002 in higher education at large, and put me in touch with the Washington based International QA in Education Group.

George always said I would never finish my doctorate because at the end of the day I was always essentially practical and did not tolerate knowledge for its own sake. Well, I think this practitioner DBA will be the making of me. It can be wholly focused on what I want and need to do as a consultant and meet the U of I's requirements. Food for thought here for Trad U.

Best wishes from us both, M&G".

These e-postcards and faxbacks continue forward @http://www.irdc.com/

E-postcard 6: A nice consultancy assignment at last …

"Fredrick. My patience is now rewarded. And it was the doctoral registration that set the folk here at IQMA thinking I was the right person to tackle it. So I am proud of myself. Not drifting away to become a student (again).

We had one of the major academic and professional management publishers at the IQMA last week. They produce, incidentally, a dozen top rated quality management journals. It was a f2f brainstorming workshop for starters on how to benchmark and improve the quality of their authors' contributions and their editors' contribution to the process. After a couple of robust hours debate, at which I was impressed with my knowledge of the recent writings on the subject (guess how!) I was volunteered to tackle the immediate benchmarking task. We resolved to let authors and editors generate the quality dimensions themselves by paired comparisons with competitive titles. It turned out I was the only one in the room who was willing to confess I did not know how to describe a high quality academic or professional article/paper.

My reticence was recently learned from seeing how ANBAR declines to accumulate its star-ings for major dimensions into a composite score. It argues that each individual will at different times want to place different values on the dimensions. Fitness for purpose, of course, but a whole host of purposes for a single article are readily apparent. Are we browsing to stay up to date, preparing a top level lecture or are we, as I so frequently am, about to visit a client with the need to have a professional consultancy assignment done expeditiously?

I also have a stroke of luck. The publisher has long since collected an excellent database of all its authors across the world now amounting to 15,000+ classified by keywords from the articles they have penned, when published and, of course, where they are employed. (Saw you in there with four entries Fredrick). I resolved to use them as my sample frame of authors worldwide to elicit what are the relevant dimensions of a high quality article and beyond that a high quality journal – and as my own unasked for afterthought, a high quality database.

All good wishes, M".

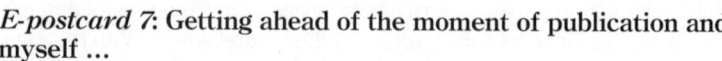

E-postcard 7: Getting ahead of the moment of publication and myself …

"*Fredrick*. More than half way through this sabbatical; seems impossible. How time has flown. We've only used the on-line recipes at the suite half a dozen times. Wonder what that tells us? And we scarcely watch the satellite movies. Prefer to walk, frankly, marvellous air quality and parks here and its only a 35 minutes MTR ride to the countryside. Seems a million miles away from the Information Society out there. And the books are still getting very well thumbed.

Anyway, my work on what constitutes a high quality article/journal/ database really is progressing well. Responses are flowing in at the speed of light from e-mails. Now I am getting ahead of myself as well and I have George's enthusiastic support.

All authors begin their articles with some form of rough draft and then maybe a pre-print stage for serious professional comment among a small circle or personal network. Once those comments are incorporated the paper goes off for review to a journal. It can and often does take well over a year from submission to publication, and the rejection rate can be as high as 60 per cent. That seems an incredibly wasteful process and certainly, I am informed by George, discourages many from even bothering with it. He says much potential scholarship never gets off the ground because of the inefficiencies of the publishing process, not to mention its controlling/sometimes incestuous editor/author charmed circles.

My idea is to get the publisher to use its much wider canvas of all previous authors on the same keywords, known as their Literati Club, to seek their comments automatically. The draft/pre-print paper can be readily commented on and updated and even "scored" on whatever the relevant dimensions are by a wide circle of relevant colleagues way beyond any personal network and ignorant of any hierarchies in place. Any pre-print when amended that beats a given score can then be automatically accepted. One added advantage of this, George suggests is that a citation index of articles in progress can be available and searchable as well as the listings of those that have been finalized. Plagiarism can be avoided by this approach too (Wills, 1997).

The publisher is over the moon with the idea. It fits wholly into its latest technology of continuous publishing, would you believe, by going back upstream. Did you know that since January this year, every time an article is accepted for one of their journals it is automatically published on their Internet site, long before any paper version bound up into an issue will reach the library or your desk? But, wait for it, there's more to come.

If the librarian has indicated to the publisher the e-mail numbers of the regular users of the journal subscription (as many as they like within reason on the same campus) then as each article is continuously published they will e-mail the abstract of it to all those thus notified, saying if you want it, take password access to the PDF and download the full text. Saves all that walking to the library and indeed honours a lot of promised visits we never make. And it gets better. By providing the author's e-mail address as an on-line link, and increasingly the references lists at the end of each paper, the message of the paper can be debated forthwith and its origination studied more deeply. Rocket science you'll have to agree.

The only muted complaints are from librarians who can be heard wondering what future they might have. George reckoned they had a while yet to survive because they would have relatively little knowledge of whose e-mails to give to the publisher. Furthermore, since it would sound their death knell, sabotage or gradualism would be the order of the day. Good point I suppose.

I reckon that they cannot beat this particular vision of the future so they had better join it. Not sure how though but hope to get some ideas as the consultancy project continues.

Must go. Best wishes. M".

E-postcard 8: Central place theory and social physics – well fancy that

"*Fredrick.* Margaret is the first to admit that her enthusiasm for what she is now calling PEERNet (her high quality article procurement and continuous improvement process) has given her little time to hear out my concerns as a marketing academic about the Internet. However, last night she gave me airtime

POSTCARDS FROM
THE OTHER SIDE

These e-postcards
and faxbacks
continue forward
@http://www.
irdc.com/

and agreed I must write to you at once. They are not going to solve the issue totally and effectively here as far as we can see for some while but I am convinced it will need to be my number one concern when I get back from this study leave.

The issue rejoices under the name of "communities of interest" (Marshall *et al.,* 1995) but I prefer to call them forums. But let me start from the beginning.

As we all know at Trad U the Internet is growing like Topsy and like crazy. Everyday more and more pages are added and more servers are connected.

To make sense of it, to get maximum personal benefit from it, takes for ever even with the search engines being built and the new crawler software which takes on your identity and scavenges the web for nourishing information even as you sleep and reports back next morning or sooner if you let it.

The overwhelming conclusion is that what most of us need and want is a one stop shopping mall (Wills and Wills, 1996), a single Internet site address where we can plan a regular visit and which has there access to all we are likely to need or want to know. Just as a good shopping mall can attract us and persuade and convince us it can cater for virtually all our requirements, so too can a well constructed Information Society community of interest or forum. As in retailing, however, the mix of services and merchandize at each mall has to be good/sufficient/even inspiring.

U of I's B School here, IMC as it is known, has now resolved to joint venture with Margaret's publisher to attempt to create sooner rather than later, such a central place for middle/senior managers that is a comprehensive graduate academic cum professional forum. Its offerings will range from:

- current awareness of global issues as they happen;

- ongoing virtual conferences on key issues;

- annual reviews and trends;

- academic and professional qualification award programmes on-line;

- focused/targeted action learning forums and workshops;

- links to accredited Coolsites elsewhere on the Internet (Binns, 1996);

- archival literature access by ANBAR Electronic Intelligence; and

- any number of associated professional services via advertising.

It will also deliberately have some light hearted goings on, a satirical column or two, professional gamesmanship, competitions and the like – indeed anything and everything to give the forum buzz and make it a convincing and worthy first stop/one stop community of interest destination.

The whole forum, with its sub-forums of course, will be sponsored by the publishers' 140 existing journals, by the B School here at U of I, and by the leading abstracting services covering the full breadth of knowledge and information published, not just the single albeit dominant publisher.

Frankly both Margaret and I think we should get Trad U and Relationship Consultancy Inc. to join in as sponsors too, either by being most worthy

Coolsites or as co-sponsors although we realize that last suggestion would be a big leap and we would have to do it all ourselves initially after we got back.

Truth to tell it sounds like a gigantic undertaking. Hard to envision but it makes neat and overwhelming sense to me as a benefits not products marketer all my life. Institutionalizing and operationalizing it will be tough Margaret reminds me!

The need, and she believes the want, is for a forum as its name suggests where the casually interested multitude and the keenly interested few, can gather together to discuss, debate, listen within a structured framework that *inter alia* captures the best of the discussions as they proceed and records it in the literature. It would all be fun surely, but relatively unsatisfactory in the Information Society, if the great debates evaporated and were never distilled as Rapporteurs notes/proceedings/articles and the like. The lack of this is one of the main criticisms heard of the Internet until recently with frequently cited evidence of quoted references simply no longer being available as servers shed them on a timed delete routine. For electronic publishing it seems that several of the seventeenth and eighteenth century approaches to the capture of discourse need a thorough rehabilitation. The forums will be able to become global, professional and learned societies.

Anyway, Fredrick, we're both converts here. G".

FaxBack 1 from Trad U: **You're underestimating us back here**

"*George*. I've thoroughly enjoyed all three-postcards you've been sending. The fact you have no response thus far has not I trust been dispiriting. But E-PC 8 needed a quick response. You left just before a couple of highly relevant publications surfaced here at Trad U and the campus has been chattering about them extensively but without much sense of purpose.

(1) The European Union's DG XIII/E reported on "Strategic developments for the European publishing industry towards the year 2000" (1996). It was adamant that electronic publishing is the pace setter for quality on the way to the information society and that to be sceptical is potentially ruinous. But the democratization of knowledge and information by electronic publishing offers great scope for broadening brand definition and in particular the building of communities of interest.

Most of us here dismissed it over coffee as the normal hype to sell more technological kit but your e-postcard did create a stir or I had better say murmur. Got the feeling several felt they were one up on you at last! (You won't forgive me, but I have been passing snippets around to the good and the bad here to seek to trigger any response and to prepare the ground for your re-entry here.)

These e-postcards
and faxbacks
continue forward
@http://www.
irdc.com/

(2) Now Harvard B School Press has published a new book, *Net Gain*, authored by a couple of McKinsey managers called John Hagel III and Arthur Armstrong (1997). It's about nothing else but communities of interest/virtual communities.

Neither the EU nor the McKinsey managers have given convincing illustrations but if what you say from Internetica and the EU and McKinsey here are anywhere near the future in store, it won't be long before we hear the details and feel the impact at Trad U. So all power to your elbow. It rings bells with me and my work behaviours and preferences.

You know how I hate spending more than an hour a day at my PC with e-mails or whatever. The only way I am ever going to come to terms with the Internet and be an effective participant in the Information Society is along the lines being mooted now. The whole pattern has got to become sensitive to me, intelligent enough to sense and sort my needs and wants. My use of the Internet for generalized and specific knowledge and information will have to be conveniently managed from a single site albeit it with myriad hyperlinks out and back.

If I can then log on two or three times each week for an hour at most, see what's what, pick up the drift and feel confident I am in the know, that's a real boon. It's worth paying a serious subscription for it. And when I need to do some heavy researching or archival browsing, I would hope the same gateway can be used to go off and explore just like I can in the library stack but that much more effectively and indeed speedily.

Currently much of the time I spend on the Internet is frustrating because of technological glitches and poor merchandizing/presentational and retailing skills. But that will come right. Then the true premium will be paid by customers (that's me) for focused and convenient resourcing which is supplemented by scope for imaginative and serendipitous linking.

The task of managing such a service seems to be that of a "meta-librarian" almost. And I would comment that may be the sensible direction for Margaret's perplexed librarians mentioned in E-PC 7. The best of them always did such things anyway for faculty and to a lesser extent students, as they got to know us all as individuals and our profiled interest areas. They could never hope to serve everyone like that but the new technologies make it a real possibility.

A lot of them here seem to be fighting the whole thing rather than grabbing it like you have. But some are brave and willing to give it a go. So, as you can see, I think you are on to something. I shall try between now and your return to see whom we can target to work with you. You may well have found the peg on which to hang it all. No hope of proceeding on all fronts at once. It simply scares us all, me included. If you can have and hold the big picture, let us mortals just do some painting by numbers.

Take care both of you. All the best. Fredrick".

E-postcard 9: European Union and net gain quite right

"*Fredrick*. Your FaxBack much appreciated here. In the enthusiasm to move on and upwards here we don't always get the opportunity for much good conceptual and philosophical thinking. I ordered both documents from the on-line bookshop, and got them in three days.

Margaret was particularly impressed by the notion that "electronic publishing is the pacesetter for quality on the way to the Information Society". I have heard her use it three times already. It's given her a direct platform for her routinization actions on her quality management consultancy project. It had never quite dawned on us that publishing can be defined as the disciplinary/ quality assurance process in an open, i.e. not censored, information industry. Those who own publishing houses are accordingly a force for setting standards and for driving them to continuous improvements.

The fact that with the Internet any and everybody can hang up data or information is not really a great help unless an accepted, non-suffocating discipline is present or somehow engineered into the situation.

Once that disciplined framework is in place, the opportunity for evaluation right across the board of the quality of what is written and how it is disseminated is massive. ANBAR Electronic Intelligence is using its individual article star-ings system to assist the searching of the literature to also build rank order profiles of the major journals on a moving average basis. Once this is easily available to Deans, promotions committees and research funding agencies all manner of competition will open up to gain the endorsement of what are more objectively perceived as the "best" journals. None of us anticipates with the arrival of continuous publishing that these "best" journals will want to stay locked in 100 per cent to paper dissemination. Indeed the more progressive are already beginning to offer multimedia multiple access within the same subscription service.

Thanks again for the FaxBack, best wishes, G".

POSTCARDS FROM
THE OTHER SIDE

E-postcard 10: PEERNet gets closer to home for authors

"*Fredrick.* I saw George's most recent card to you and wanted to bring you further up to date on how my PEERNet project is progressing. Its brought me very close to the abstractors in the academic and professional literature. I have something quite astonishing to report that builds on, but I do not think contradicts, what George had to say.

It is the abstractors' growing opinion that the quality of the abstract writing and the consistency of their quality ascription's will be the drivers on the demand side of knowledge markets into the next century. They believe their branding (as with good food and hotel guides and the like) will be more significant than the individual journal within which the article originally appeared. I would emphasize that I am talking demand not supply side market here. But if they are correct, the journal brand can be expected to disappear as a retail brand but to survive as the accrediting brand for the original author's work. Am I making myself clear on this?

So what am I benchmarking journals for *per se*? Why not either articles which must each stand on their own feet or the new intelligence services on offer? The best intelligence service for readers will be that which offers the fastest, most intelligent access to the very best articles in the world.

These e-postcards
and faxbacks
continue forward
@http://www.
irdc.com/

This seems to be where ANBAR Electronic Intelligence has been headed for the past five years, and they appear to have got it right now for managers at large, management academics, computer specialists and civil engineers. And once that is totally networked at Trad U we'll be in business with a vengeance.
As ever, M"

E-postcard 11: The Dean is determined to launch an e-journal from U of I

"*Fredrick.* The Dean here at U of I has decided she wants to launch a new Graduate School Proceedings/Journal on the Internet. She sees it partly as PR for IMC but also to ensure they capture a lot of good "grey" literature, good discourse in various workshops, seminars and colloquia.

As my swan song here as Visiting Norbert Weiner Professor of Marketing she asked me to do some thinking and lead a taskforce to get it started. She is convinced the technology is not the issue. Easy, she says, once you have the hang of it. She wants me to start with a review of the existing Design and Process Monograph Series IMC has produced for a decade (many of which have just appeared in print edited by Alan Mumford from Gower Press, 1997) and build from there.

The challenges have emerged thus far as follows:

(1) Giving some coherence to the disparate themes that have emerged. To impose a theme would surely miss the point.

(2) Ensuring something like a smooth/regular flow of material. This has not been so in the past.

(3) Convincing Faculty that the new medium is a worthy/creditable vehicle for publication. The Dean is, I think, rightly convinced that unless this is achieved the idea will be still born or a waste of time.

So, the myth of scholarly skywriting as the new future is once again not supported (Kling and Covey, 1995; Hitchcock *et al.*, 1996). Rather, we are exploring how we can make sure it is included in the ANBAR Electronic Intelligence service. Our thinking is that if we can ensure the contributions are there and rated well, then the Proceedings will be a more than satisfactory primary medium for publication. Fortunately conversations with ANBAR's editorial selection panel show they are eclectic in their coverage, deliberately including articles of greater benefit to consultants than to academics and vice versa.

They have however also given us an unexpected piece of good advice. Margaret's publisher where she is consulting has for two years now hosted an Internet Free Press (IFP) as a co-operative of people/institutions like U of I seeking to launch e-journals. For a small annual subscription IFP offers advice and fellowship on how to get the e-journal up and running, access to a subscription service if required, enduring server space/archive for every article published, abstracting into the ANBAR Electronic Intelligence services, plus a

POSTCARDS FROM
THE OTHER SIDE

golden paper copy master lodged at the British Library for belt and braces (Butler, 1995; Mathieu and Woodward, 1995).

I found it difficult to understand the logic of this but apparently its an off the wall strategy to run with the revolution, on the streets as they put it. Came out of a think tank on embracing (Wills, 1995) rather than resisting electronic publishing they had way back and is paying handsome intellectual dividends. They simply believe that as with paper journals of old, enthusiasm will wax and wane for e-journals and they will be there to take over the professional support of any title that wants it. Their own origins in paper 30+ years ago were on exactly that basis. Since we are on the upswing not the downswing it's no threat but it's pleasant to know its there if/when we stumble.

So, no peace for the wicked here. Ciao G".

E-postcard 12: **What can we hope to transfer back for Margaret and Trad U?**

"Well. Fredrick our last e-postcard! We have not forgotten that the terms of our year out here in Internetica were that we had to come back with two action plans, one for Trad U and the second for Relationship Consultancy Inc. Nor I imagine have you, and we have been dropping hints already!

POSTCARDS FROM
THE OTHER SIDE

First, Margaret is determined to finish her practitioner DBA and has agreed with the Dean here, who is her supervisor, that the development of the PEERNet concept shall be the focus, linked as it is with procuring quality articles for publication. She wants to use us at Trad U as well as the Faculty at U of I as prototype campus based Authors' Clubs sponsored by the Literati Club itself (Wills and Wiles, 1996). Sounds off-the-wall but with the priority these days to do more and better research, it could well ring the right bell. There are one or two prototypes already in Malaysia and the UK so we won't be 100 per cent guinea pigs.

For her Partnership consultancy she foresees a wave of workshops with authors, both from academe and from corporate enterprises both on how to contribute towards and gain benefit from authoring, and also how to excel at abstracting/extracting of intelligence from the Internet. Here she expects to be deeply involved in the population of Intranet systems in corporates, as well as making specific use of the QualNet Forum the publisher already offers and ANBAR Electronic Intelligence services in quality *per se*.

For George, the smartest plan discussed was to request a sabbatical funded this time by U of I to give him time to digest it all and make sure the IFP Proceedings journal from IMC is effectively launched. But without too much reflection we agreed the real challenge was at Trad U. So, it's going to be an all out effort to get communities of interest and forums going from the Trad U campus. We have talked the publisher here into him becoming what they call the Convenor-in-Chief for Marketing and Logistics. The Dean at U of I is going to work along with him too. All the editors of titles clustered in marketing and logistics are being invited to be regular convenors, running a chat line/authors feedback/conferencing service as an adjunct to the regular journal

These e-postcards
and faxbacks
continue forward
@http://www.
irdc.com/

publication. George has the further remit to find convenors at Trad U if they are willing to stand up and be counted for human resource management, Strategic Management and of all things Asia Pacific management. Whether that is expecting too much too soon at Trad U we would not know but you did say in your FaxBack that some of our colleagues are looking livelier and anxious to score a point or two. Internet fame beckons!

To assist it all along, the publisher runs an Internet Convenors' Forum for all Convenors, would you believe. The equivalent of the IFP for the e-journals so to speak. Anyway, we have both joined in its discussions of late and it functions very well as a support mechanism with occasional f2f as well. There was a big meeting here for two days in fact last January. And the publishers themselves give a great deal of routine support, simply picking up the gurus, so to speak, for vital but clearly defined work only. They are sufficiently realistic about the stereotypical academic's adherence to schedules to have a piece work payment rate too: no piece on time, no payment; three misses and you're off the team. I think we already subscribe to *Internet Research* journal, which they publish both electronically and on paper, and they summarize the *Proceedings of the Convenors' Forum* there from time to time as well as dissertation projects from within their own Internet Research and Development Department. They have produced some ground breaking stuff on evaluating virtual conferences (Heal, 1996) and Coolsites (Binns, 1996) there too. They are great believers in the learning organization model of management (Reeves, 1995).

Finally, Fredrick, I am determined to put all my own programmes on the Internet linked with the ANBAR Electronic Intelligence services. I shall run it in parallel with the taught courses but expect the team in continuing education to be more than happy to use it for their distance learning students and to incorporate the whole Web Forum technology just like U of I. The beauty of it all is there is virtually no development cost and Margaret and I have both picked up enough knowledge of html to be able to set our offsiders to work with the course materials disks as they stand.

So, both of us have decided to return without any grand plan for what the Partnership or Trad U at large should do but a determination to act as converts and to welcome help from anyone who wants to get stuck in. Any volunteers, Fredrick, apart from you that is?

Greatly looking forward to getting away from these on-line recipes and down the road for a Trad curry. See you as soon as we can, M & G".

***Faxback 2 The George Report to Business School colleagues at Trad U:* Reflection on what has been and a picture of what can be at Trad U as a great scholarly institution**
Centuries of stony sleep
Scholarly publishing as an industry has scarcely changed in nature for hundreds of years. Although innovations in typesetting, papermaking, inks and despatch systems have incrementally altered the way things are done, the things themselves pretty well stayed the same.

The story of the university industry has uncanny parallels to that of learned publishing. They are indeed bedfellows, sometimes uneasy ones, and without the one the other would hardly exist. Universities first came to be in the Middle Ages and again have hardly changed in form or function for 500 years.

Initiatives such as the British Open University did provide a change in the notion of what a university is and does, 30 or so years ago. For the first time university education was democratized into "open learning". The Open University (OU) was one of the great additions to our social infrastructure – a government-underwritten, open-access university where if you were 60 years old and taking care of the house in a village in Scotland, and fancied taking a degree course in the History of Philosophy – you could. All you did was paid your (modest) fee, registered, and followed a system of TV-transmitted tutorials (sent out at off-peak hours) plus worked your way through set books and course materials, plus attended optional summer school. The OU set a global pattern many others have since emulated.

It changed the dynamics of education from an exclusive paradigm to an inclusive one. The exclusive paradigm said that if you were rich enough to travel to one of the great seats of learning and maintain yourself there for several years while you studied, and you were deemed worthy of being there, you could go. Capacity was small, limited by the campus and the number of suitable teachers – and the need to maintain exclusivity of course. The OU and democratized open learning offered a theoretically limitless capacity – through the medium of television one teacher could lecture to a million students simultaneously.

POSTCARDS FROM
THE OTHER SIDE

The democratization of higher education led to a need, still largely unfulfilled, for democratization of scholarly publishing. Instead of full-time scholars poring over esoteric tracts, consumers became people who had many other things in their lives, and people with vastly varying degrees of prior knowledge. The exclusive publishing paradigm, mirroring the universities which symbiotically existed alongside them, said that if you were clever enough to understand the stuff and able to gain access to the library to find it, you could study it. Democratized open learning threw out a challenge that said – without "dumbing down", let us present information we can learn from in a manner and in media accessible to ordinary people.

At the back of these two great industries, and in a way the glue which binds them, is a third one – the library. Again, just as a publisher and a professor from 1497 would have some familiarity with the roles of publisher and professor in 1997, so a 500-year-old librarian would be able to walk into most libraries in the world (most particularly academic libraries) and be on familiar turf.

This potted history of our great institutional trinity of learning – the scholarly publisher, the university and the library – is provided as a counterpoint for what is to come next. For here at the end of a millennium which spawned these institutions, we are standing on the edge of the most profound change in their entire history. One more profound than the printing press or the Open University's democratization of learning, as it affects all three simultaneously and, literally, earth-shakingly. So much so that the publisher,

These e-postcards
and faxbacks
continue forward
@http://www.
irdc.com/

librarian and professor of just a few years ago would have more in common with his or her counterpart from half a millennium away in the past than with those of a few years hence. Everything has changed, changed utterly, as the great prophetic poet WB Yeats put it.

As we stand on the brink, with a foot in the old world and a foot in the new, the challenges for those of us in these three institutions are frightening in their profundity. We can no more stop the metaphorical tide than Canute could. But our actions and ideas may shape its flow, if we choose to try. If we choose not to, our impact will have been as footsteps in the sand, washed away by change. But the opportunity is to leave a legacy which may shape our great scholarly trinity for the future, just so little, but profoundly so.

That is why we believe we must grasp and grapple with the vision, and seek to leverage our wisdom, such as it is, to embrace the future. Otherwise we will have, quite literally, left no mark.

What we can and must do now
Our business must be re-thought. We cannot see tomorrow's business with yesterday's eyes. It is time to separate what we do from what we are, and to be again what we are. What we do and have done is no longer appropriate as the driving impetus of the business. We must find new things to do, which reflect more accurately what we are and what we need to be.

Our proposition is that our customers, now and in the future, will increasingly seek an interactive relationship with the knowledge they consume, and that this will increasingly be perceived as more satisfactory than the one-way relationships they currently have with publishers, universities and libraries. It is on that proposition that we will try to point the way forwards in tangible terms.

1. *Virtual communities:* We must create virtual communities around areas of current and future activity, led by us but shaped and moulded (and populated) by editors, editorial advisors, past and potential future authors. The most valuable currency we will have to trade is knowledge. We must therefore concentrate on the suppliers and manufacturers of knowledge, and ensure that the supply remains fresh and plentiful. We can do that by:

(1) Providing the wherewithal for these virtual communities to exist, via well-designed Internet discussion fora, the purpose of which is to aid the creation and sharing of new items of knowledge

(2) Understanding the motivations to publish and the difficulties of so doing, and ensuring that we are providing acceptable validation systems (e.g. properly conducted community review, listings in various indices etc.); with help and guidance in crafting an article; and that our editors are effectively mentored and managed to be able to build relationships and referrals with authors, and perform to acceptable service levels.

(3) Providing guided access to our databases for research and data gathering for our authors. We have a structure (e.g. Literati Club) which can be further leveraged. If we really believe our knowledge suppliers to

be of paramount importance to our continued existence, we must treat them as first class passengers while they are freely giving of their valuable commodity to us.

(4) We must further encourage the supply of valuable knowledge materials from the university structures we manage and influence. We should design patterns that will "harvest" the outcomes from action learning and action research programmes in publishable form rather than hope they turn out that way, and we should encourage our partner institutions and others besides to do similarly. We should coach our students in creating publishable outputs. Much real learning and insight is lost to the wider community because it fails to be captured. It gets lost in what we normally term the "grey" literature. We can hypothesize that tangible output through publication would be good for the university in improving students' post-programme appreciation and loyalty, as well as acting as an effective publicity channel for the university and its courses of study. All programmes should be designed into and around the virtual "knowledge manufacturing" communities described above. (We have the prototype in Project Harvest already.)

These e-postcards and faxbacks continue forward @http://www. irdc.com/

2. Brand management: We have carefully created academic brands over many years. We must devote much strategic brain power to their function in tomorrow's world. Whether we should abandon them as no longer relevant, manage them to accrue short-term gain, or leverage them into an e-presence is a strategic decision which cannot take shape by default and neglect. It is clear

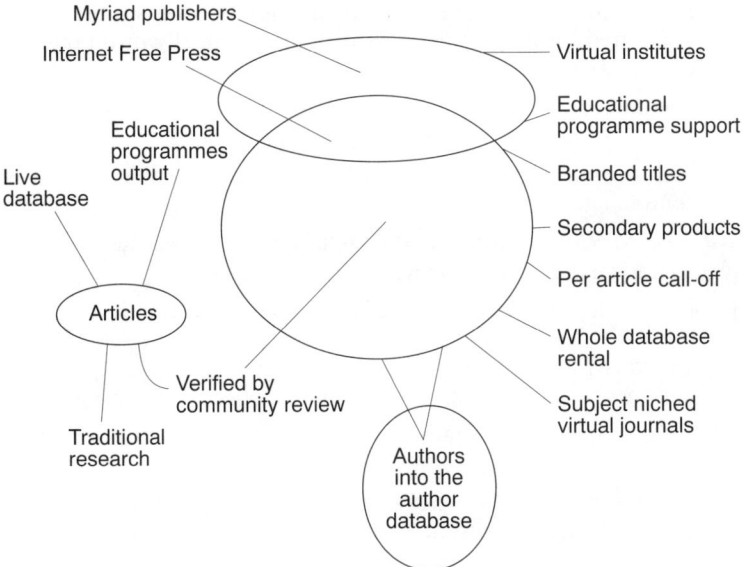

Figure 1.
The Database
(keyworded, etc)

that publishing will sooner rather than later lose coherence if it continues to be driven around the brand titles as they are now, just as it will if it built solely around the annual advance subscription as the sole unit of income. That will not be how our customers will see the world, and that will not be how our farsighted competitors will act. Our databases are – we now see only too well – more than an agglomeration of brands and cannot be managed and sold for long in such a manner.

3. *Database management:* Our database of articles (presuming we limit ourselves to articles and their derivatives) will prove to be the mainspring of the knowledge/ learning business if it is to survive and prosper. We have discussed inflows above, and made reference to a change of emphasis from traditional journal brands to the database.

The database must spawn products and services – as many as we can create, as their creation and distribution is an incremental expenditure only after they have been assimilated into our database. Indeed, we have no need to create them at all – we can give our customers the tools to create them on their own.

Examples are:

(1) Traditional journal brands, albeit supported by a web of communities of interest around them.

(2) Access by keyword search to articles, charged per access, per download, charged by monthly flat fee, or any or all of these combinations. We must find efficient and economical means to deal with multiplicity of access charging and collection and not be hindered by present lack of ability or reluctance to do so.

(3). Access to subject or industry niched areas, selected by us or auto-generated for a customer, supplied electronically or on paper or by combination, as "virtual journals".

(4) Secondary products derived from the database, in a multiplicity of forms, as brand spin-offs or as v-journal support.

(5) Institutional subscriptions initially to the whole database, followed by a continuing top-up subscription.

4. *Database enrichment:* Not only do publishers trade their own products, but they also trade those of other knowledge-gatherers.

(1) Already a nursery for e-journals has been created in the Internet Free-Press which will become, through investment in promotion and publicity, a further means of both capturing raw materials and exploring brand parameters and innovative publishing practice in the e-world.

(2) We also trade other materials through Anbar Electronic Intelligence albeit reliant on paper for much delivery still.

(3) We must seek other ways to add to our collection of knowledge at low or no cost. Copyright ownership of electronically disseminated materials is emerging as a key debate. It is likely that more authors will seek to retain

their own copyright. We must swim with this tide rather than against it by encouraging authors to retain personal copyright and accrue royalty income from accesses from our database as a radical "New Deal" with scholarly authors.

(4) Knowledge also arrives and is greatly facilitated from other directions – at conferences, lectures, seminars, consultancy interventions. Wherever knowledge is created and captured – we can and should be there, helping if needs be with creation and capture, trading it ourselves.

5. *The new library:* Our databases are designed with library tools, and must continue in an ever-simpler, more cost-effective and more accessible manner. We have started the process very effectively with keywords, typology, quality ratings. We must both validate and adjust our systems and continue to think from a learner's and researcher's point of view in accessing our databases. We have taken over much of the function of the library and librarian, and our current and future relationships with these institutions and people must be carefully and properly thought through.

> These e-postcards and faxbacks continue forward @http://www. irdc.com/

6. *The new university:* Our databases are already providing the impetus for an extension to democratized open learning, supported by e-mail and Web forum discussions, e-tutored globally by (often) the same authors who write for us and recommend our products to their institutions.

7. *Virtual institutions:* The knowledge and network of people we are creating can be formed into entities some of which we are still to conceive of. However, it seems clear that they will engulf both the institutions of higher learning and the professions as we know them now.

Conclusion – a new way
None of this is beyond us. Much of it we are doing now. What we must do is seize the moment intellectually and indeed as social or private investors, before it passes us by. Our organizations will be like fish out of water if we think of them and run them on pre-electronic lines in the electronic era. Much as we must of course engage in tangible activities towards future re-shaping, we must do something more difficult and more challenging than that. We must learn to think in a new way. Can we?

George wishes to record his thanks to John and Peter at U of Internorbert, Internetica, for crystallizing the issues reported here in discussion with him and Margaret before they returned home to Trad University.

Faxback 3 Static and dynamic quality assurance in the Internet age: Is there any comfort to be had?

You asked, I promised, so here are my observations and conclusions as the result of my time away in Internetica with George.

When reflecting on the QA challenges for our organizations in the Internet age I found myself engaging in something closer to a philosophy seminar than

one on which quality tools and techniques we should adopt. But then, we must understand the philosophical base from which we proceed.

As a quality assurance pragmatist that makes me both comfortable and uncomfortable. Comfortable because I know from long experience that to start with operational rushing around usually ends in strategic tears. More haste, less speed, as my mother used to say. Uncomfortable because of a propensity for action which I increasingly find means more than pontification, debate, meetings about meetings, memoranda about memoranda. Let me see if I can reach equilibrium among these forces of discomfort; discuss for a moment two other competing forces which need to find equilibrium; then move on to some recommendations as to what can be done. I say "can", but I think I will find myself saying "must" as I go on.

Ever since the beginning of rational thought, humans have sought to create, and to preserve. These are the great metaphorical, spiritual complementarities and opposites; the yin and the yang. Creation says – what we have is not good enough. We must change it, destroy it if necessary. Preservation says – yes, of course it's good enough; it's worked fine for all these years, hasn't it? Sometimes the preservers win, for a while. Sometimes the creators win, for a while. And when the forces of creation prevail, as they inevitably do, then the preservers move back in and say – thank you very much, that was exactly what we needed, now we will make sense of it all. And so the struggle, the cycle, begins again. Those who create and those who preserve are naturally antithetical (look at revolutions) but necessary. Without preservation there would be only anarchy; without creation there would be only stagnation and death. The American writer and philosopher Robert Pirsig calls these two forces the static and the dynamic.

See for example: *http://webpages.marshall.edu/~striz 1/quality/*

In seeking to make sense of the electronic revolution we can see static and dynamic systems in action clearly. I only present the philosophy lesson to see if we can distance ourselves from the day to day excitement and reach some kind of strategic high ground.

Assuring the dynamic

Without dynamic forces which irritate, destroy, create dis-organization, expense and trouble, our organizations will die. People able to harness dynamic forces are hard to manage, sometimes unpopular, annoying. Yet we must nurture them.

The forces which create and those which preserve are not just different; they are antithetical. We can manage creativity right out of our organizations all too easily; we can impose too-heavily structured systems; we can squeeze people's time and space. It has been called the corporate immune syndrome. Effectively we motivate creators to conform or get out. Some, exhausted, will conform; the rest will get out. Then we are lost, at the mercy of crisis as a change driver.

Quality assuring the dynamic is as important managerially as quality assuring the static. We have no need and no excuse any more to over-centralize. We can and must work with creative people across cultures and geographical distances; we should look to manage our dynamic forces outside of our static centralities. As I sometimes remind people, and I sometimes need to be reminded; we create and manage systems; they don't create and manage us unless we allow them to.

I heard a story about the Sony Corporation. Sony have three teams working on each of their products. One team works on incremental improvements – how to make what is just that little bit better. One team works on support systems and products – how to support what is a little better too. A third team works on actually making what is obsolete, by looking at customer usage, and at changing trends. They are paid to think and to play. But Sony is a large corporation and most of ours are much smaller. Should we pay people to think and to play; to destroy what is by creating something new? Can we afford to do it, and can we afford not to? When things change, as they are changing now, it is not a rhetorical question, but one we must ask with due seriousness.

Assuring the static
Without static forces, which are seen as stultifying, bureaucratic, dead-weight wet blankets on our creative fires, our organizations will never be able to move beyond fleeting glimpses of good ideas. People able to harness static forces can be cold, unimaginative, slow, negative and resistant to change. Yet we must nurture them.

We have few excuses for not assuring the reliability and managing the variability of what we can, easily, assure. Systems such as ISO 9000 are well-documented, low-risk and readily implementable. Bringing disciplines to areas which respond to discipline isn't just good for business. It's good for our own sanity and self-respect too. Most people don't want to do waste work or sloppy work. If we don't assure the assurable, we will leave little space for creativity, and for managing the difficult psychological task of seeking equilibrium between the static and the dynamic.

There is nothing intrinsically dynamic about the Internet; about e-publishing, or e-education. Some parts of the process of design, invention, human interaction, or just simply thinking hard, are dynamic and will best be managed outwith our organizational controls, yet linked to them. But many of our electronic activities are routinizable, and can fall readily under routine assurance systems.

The principles of quality assurance are simple; understand correct specifications and produce to them; understand causes of variation from specification and control them, prevent them from causing random difficulties; leave an audit trail and use audit data to improve performance. We can do all this creatively and humanely.

These e-postcards and faxbacks continue forward @http://www. irdc.com/

Buzzing and Humming; quality assuring Internet activities
I wrote previously (*e-Postcard 3*) that there is "the hope of getting cost-effective linkages and lateral thoughts. But they are not overly optimistic about great results because of the earlier view that a lot of attendees do not want to contribute. They just want to visit and look and maybe learn.. and why not?

Nothing new in that at all but we all wonder whether the Internet has got some ingredient that can set a new pattern alight. It's going to require some lateral thinking. The analogy they make here is that an important part of any traditional conference is face-to-face in the bar, lobby and the like…"

An exchange of e-mail is not the same as a conversation leaning on a bar or over coffee in a hotel lobby. There are no social dynamics; no body language, no tone, no eye contact. Interaction can of course take place, build and grow, but on an intellectual, analytical level closer to a Socratic dialectic than a spontaneous chat. Internet buzz isn't a replication of face to face, just as video-conferencing isn't, just as a telephone conference call isn't. It is what it is.

It may be that assuring the interactive, conversational buzz requires similar disciplines to assuring the archival hum; archiving, search tools, managed links. In searching for answers, we must look in the right places.

Next steps
Over the next two years we need to seek stable anchors in times of change. Our systems can and must become impeccable, with waste work and waste effort managed down to minimal levels. We know how to do this, and it is time to grasp the nettle and do it. We need to devote high-level thought and debate to the scope of a QA system certified to ISO 9000 standards, which is especially designed to take on our Internet activities as well as our more familiar ones; our dynamics as well as the statics we are more familiar with. Having done so, in depth but not at length, we need to be making our systems shipshape; writing specifications, managing to them, controlling (not necessarily eliminating) variances, auditing the activity, improving based on the data we collect.

As found in their work with IMC's QA systems, we must think of ourselves as service providers, rather than as controllers of everyone we encounter, and establish systems to assure the quality of our services to customers and, increasingly, suppliers. It is from this perspective that we should address our voyage towards ISO 9000 and beyond.

At the same time, without wishing to be overly paradoxical, we must seek to destabilize what we are doing, by surrounding ourselves with people able to think outside of our own limits. We need to be working with innovative, top-class thinkers among our own staff, our suppliers and our customers; to be rewarding and encouraging (not managing and controlling) them, and leveraging the capability of the Internet for remote meetings and brainstorming. Those among our own people who we believe to be radical, dynamic thinkers, we should deliberately distance from the organization and force them into using the technology to its fullest extent, whether on temporary sabbatical or permanent out-placement.

In our use of the dynamic, particularly on the Internet, we should be focusing on our skills in information capture and access provision; on our skills as archivists, librarians and added value publishers. The more we take away creativity from routinizable activities, the more space we leave for creativity to flower where it needs to.

George says that conclusion is one he reached as a young academic 30 years ago in relation to marketing and sales activities. And he says he picked it up from the balancing of production management and research and development. So it seems to be a universal phenomenon alright.

Our action plan therefore seems straightforward to conceptualize! We need to nail down the jelly when we feel its going round in circles and let the fascinating flowers bloom. So where's the jelly? Where are the fascinating flowers? And are people with the congruent work preferences at work on each, feeling comfortable and understood?

We have got far enough now with the Internet to start managing it professionally just like everything else. No need to indulge the lads any longer. Lets look for the business models and the roi. That's what to tell our clients and to show them how.

Margaret

Margaret wishes to thank John and Peter at IQMA, Internetico, for the advice and guidance given in drafting this report.

E-postcard 13: I dreamt I was a global tutor

"*Donna*. I trust the Dean's job is as enjoyable still as when Margaret and I left U of Internetica. I'm sorry it's so long since we were last in touch but I have an alibi you'll like. Here at Trad U we have had the most amazing developments. Can scarcely believe it. We knew Fredrick, the former Dean here and a great family friend, had been passing our E-postcards and FaxBacks around, but at this month's meeting Trad U's B School faculty was buzzing with revolutionary talk! Some of the least likely faculty here were arguing that we should take an "offensive/defensive" strategic position on the future of the Internet in education and training. Sounds good!

POSTCARDS FROM
THE OTHER SIDE

It transpired that two of them had been approached by their text book publishers for the electronic rights. Under these arrangements their titles would be made available (on the far side of a credit card actuated barrier) chapter by chapter on the Internet. But what really set them buzzing was that they had also been invited to design an MBA core course framework to go with their texts to be offered by one of the top accredited distance learning B Schools in the world. If they were willing to offer tutorial feedback to those taking such programmes and using interactive Web Forums, an astonishing \$US250 fee per student was available (Sounds like the Virtual Institution we discussed in FaxBack 2, Figure 1 is here already).

So the debate that is now raging here at Trad U is whether we should tolerate/allow/even encourage faculty to undertake such tutorial activities privately or insist that they put all such undreamt of opportunities through the B School. That conjures up shades of the old debates in the 1960s about consultancy activities outwith the B School of course.

This time it looks as though we will make the right decision. The two Professors who have been asked by their publishers in this way were the first to

These e-postcards
and faxbacks
continue forward
@http://www.
irdc.com/

spot that they could not cope with large student numbers and would need the sort of support that other faculty could give anyway. But more than that, they asked why the B School itself could not undertake what the publishers seemed intent on doing? So, building on so much of what Margaret and I learnt from our time at U of I, we proposed and got immediate acceptance of the notion that an Internet Research & Development Centre should be established.

It's called the IRDC for short, and here's a further surprise: it is going to have three equity elements in it rather than hoping or waiting for a tranche of government funding. Element 1 will be use of the B School's assets and consumables, element 2 will be a straight cash offer for shares among faculty members and the third is to be what are known here as "sweat" shares – achieved for actually doing something electronic.

We have no idea what level of resources we will need and even the finance faculty are bemused at the prospect of crafting a budget for year 1. The best idea to date seems to be to take it project by project. And you won't be surprised to hear that we are going for the MBA core courses as our starting point. Margaret has already talked to ANBAR Management Intelligence and done a deal with them to keyword each module just as you have in IMC at U of I. They said we are the 9th B School now to join with them. They began working with Canadian School of Management in Toronto only last month and have new clients coming up in Dubai, Bulgaria, Switzerland and Bosnia. The ANBAR drive is led by an ex-merchant seaman with an MBA called Eric! Seems the nautical background made him comfortable riding the waves of the future.

Anyway, back to budgets. What do you think of our IRDC notion? And what budgeting processes do you use? I know that your faculty all do their Internet activities with you because that IS the mainstream work of U and I and of IMC in particular. But what experiences are there elsewhere of it being treated as an internal or an external activity by B Schools?

Sorry to ask so many questions … but we need any inputs you have to offer. It's all so new here and so ambiguous it will soon drive us crazy. The decision to go for the core MBA courses as the first activities at least means we will talk about real issues in the coming months rather than chattering while Rome burns.

Amazing, is it not, how lateral events/external enemies end up achieving progress/uniting us for a common purpose? I could have argued the case for ever at faculty meetings without much success, I suspect. One word from a publisher and the decks are cleared for action.

Greatly looking forward to your reply. Best wishes, George".

E-postcard 14: Internet R&D Centre's secret weapons

"George. It was good to hear from you and to learn that the publishing world is providing the fillip to get you on the move. That is not an unusual pattern because they are among the largest inventors with flexible strategies and an eye for potential real returns. The other major players, who you and Margaret both met when you were here, are the former library subscription agents like Ebsco,

Blackwells and Faxons/Dawson. Our advice to your colleagues is to go with the opportunities with the publishers to become global tutors from their serialised texts, but at the same time to advance as you are suggesting via Internet R&D Centre. The strength of course of this latter strategy is to get all faculty double benefits from the system – greater fame and greater fortune. Trad U is well-advised to collect a royalty or campus tax, but to allow free access to its facilities as the revolution unfolds.

We found that there are two parallel strands which have to be budgeted for – the first is finding the market out there for whatever particular services will be generated; the second is keeping up with the emergent technologies. On the second count, i.e. new technologies, my advice is do not attempt to create new ones. There are very clever computer folk out there with megabucks to spend. Put all the effort into developing the customised applications you want and which will make life much more bearable for the users.

Anbar's site is a very good example nowadays of how they have worried their way to find most elements they need. They have bought in technology from Reed Technology in Germantown, Va. and outsourced a lot of the page graphics work. Their creative marketing has gone into their League Tables and their Hall of Fame/Citations of Excellence.

One of the best ways forward for IRDC will be to have its own web site and to have a regular column there that surveys the "new technologies" that are relevant and shows how they can be used effectively. You might even put edited versions of these e-Postcard exchanges up at the IRDC site. We seem to be roving over the important implementation issues well enough to be worth others eavesdropping on us.

However, by now you may be wondering why I headed this card "secret weapons"; anyway I hope so. There are two not so secret lessons that we have learnt the hard way here in Internetica over the past couple of years.

Lesson 1 is that it takes creative/innovative folk to build a site but upholder/maintainers and controller inspectors to keep it good and populated. As such, most of the teams here have gone through a metamorphosis and are no longer quite such a gung-ho wild bunch as they were. Derek says that's because they and we are just a little bit older. It could be, but, like most husbands, I think he has missed the point. The job requires routinisation of 80 per cent of what goes on and unless that happens all our site visitors find the site in a state of too much flux. We can flux way as much as needs be, and don't get me wrong and think I am saying the creative days are gone, but it only needs to be 20 per cent that bright and new.

Lesson 2 is that regardless of how "good" the site might become, or how many Tiger Citations it might get, if the process role we now call Internet Marketor here is not fully staffed and sustained, with monthly accountabilities, the e-messages simply don't get created and linked out there at search engines and the crowds don't find their way spontaneously to the site. The undoubted exemplar here is Blanchard & Barbs work at their Asia Pacific Management Forum.

These e-postcards and faxbacks continue forward @http://www.irdc.com/

Now you might say that your initial plans are modest and that the work will be on making a better deal for your captive student market. But please be assured that the right Internet Marketor will be actively selling places/recruiting for the programmes themselves. And that also means much more extensive use of the texts colleagues have penned. At Trade U.

Hope these comments are of help as you get cracking. Best wishes to you both from Derek and I. When are we going to see you and Margaret over here again? Donna".

POSTCARDS FROM
THE OTHER SIDE

E-*postcard 15*: Emergent global columnists on the Net

"*Donna.* Your two lessons/Internet Secret Weapons in E-postcard 14 set us thinking hard here. Some of the faculty have got best selling texts that are used all over the world and others a series of titles that sell just 2,000 copies and don't go for a reprint. And again, others not at all well ... without any apparent reason for the difference. Which is meant to be my oblique way of sharing our emergent thoughts on how to create and sustain a great community of interest on the Internet.

How, we argued, did a text get and hold a global rather than a campus/country market share and how did others flare up, sell well and then pass quickly away? And via the citation analysis, what were the dynamics of word of mouth so to speak and did nothing succeed like success? Naturally we did not reach any world shattering conclusions, so we decided to review the sites that the people at MCB University Press and their associated network have created and look at what they were about.

We started with the Asia Pacific Management Forum, the exemplar you suggested. We concluded that their secret seemed to be a great blend of stolid/text book type material in their archive but good buzz via their news items/pick of the week articles and their emergent global columnists from "Blanchard's Travelogue" and the universal dilemmas and joys of "An Expatriate Asian's Diary in America" to "Boyes Asian Code Words". This latter phenomenon fascinates us enormously now. Its a logical halfway house between e-mail top of the head responses and the tough going of Internet Conferences as captured at the Virtual Conference Centre. Its a great format which we think can act as a useful magnet to bring interested folk back regularly. It's en route to create a breed of Global Columnists.

We went off and looked around the other MCB University Press Global Columns and reckoned that they are on to something very powerful. By using these columns as a frequently amusing/anecdotal but serious commentary point on their forums, the whole thing becomes more than just a library-like site. They must have done a lot of hard thinking about how to make their sites more than the virtual equivalent of a library stack selection.

It would be invidious to choose between them but for all the faculty "Write Stuff" from Robert Brown in Australia was so close to home it was bound to be a hit. He uses humour, wry comment, whimsy, dramatic irony, as well as jokes which seldom travel too well between cultures, with great effect. He's obviously been at work some considerable while.

Some of the more recent columns are equally fascinating, with their own whimsical positioning, like Mungo Park's focus on global marketing and Burke and Wills writing on logistics. They both have considerable warmth and seem unafraid of intimate comment with which quite a few readers might well disagree! Tales from the Wild Frontier seems to be an interesting alternative angle which has less about quality management than we expected but reads well. On balance we felt a stronger albeit whimsical connection with the subject is likely to be more powerful but how knows yet? I suppose Hit-meter tracking technology will soon tell them all.

Some of us have resolved to volunteer to have a go at becoming columnists on the wave of additional forums MCB University Press has just launched and reports it will dynamically populate with ISO 9002 disciplines in place by January 1998. They gave us the full list and I'm enclosing it here, just in case any of your colleagues at U of Internorbert want to volunteer too. They don't pay a fee by the way! Their message is that they give us a Global Soapbox, good Notes of Guidance from the Internet R&D Centre, and finally they throw in archive searching gratis ... which truly is a great bonus. The bad news is if you miss the ISO 9002 deadlines you get sacked at once and that's why they keep the copyright on the pseudonym for the column and I suppose you can't blame them for that.

The full list of MCB University Press sponsored forums wanting either their first or additional (they say they can't see any limitations yet!) Global Columnists are as follows:

- Africa Regional Forum
- Airport Business Forum
- Australasia and South Pacific Regional Forum
- China Regional Forum
- FINA Petrochemical Industry Forum
- Financial Management & Services Global Forum
- Global Education Forum
- Global Forum for Police Studies
- Global Healthcare and Environment Forum
- Hospitality & Tourism Global Forum
- Human Resource Network
- Information Management Global Forum
- Internet Free-Press
- Internet Research & Development Centre
- The Library Link Forum

E-postcards from the other side

39

POSTCARDS FROM THE OTHER SIDE

These e-postcards and faxbacks continue forward @http://www. irdc.com/

- Marketing and Logistics Forum
- Operations and Production Global Forum
- Property Management and Facilities Global Forum
- Qualnet
- TopMan Global Forum
- The Virtual Conference Centre

They are an extremely broad church at MCB University Press as you can see, and they are all at different stages of development, but that's life on the Internet. Emergent itself, with none of us truly sure what will or will not work well.

Let's have your own thoughts when you have time. The IMC site you maintain is already a world leader for B Schools of course. What are your development plans for it? I see you have just linked up with the Canadian School of Management too. I suspect that putting them up on the Internet in six short months from a standing start might well have lessons for us here as we trundle forward.

Best wishes from Margaret; as ever, George".

POSTCARDS FROM
THE OTHER SIDE

E-postcard 16: **Emergent master teachers on the Internet**
"George. You got to me with your Emergent Global Columnists last month. That was a postcard that set our bars rattling over here. I had not realised just how far MCB UP had got with their forums. There are still a few missing links I saw but the use of ISO 9002 disciplines looks like an excellent way to be sure the frequent visitor gets the urge to pop back and see what else has happened. And congratulations to that Daniel Newman who crystallised the Guidance for Internet Columnists.

Anyway, on to the new order here now. We've instituted a new grade of faculty member and give salary bonuses for top ratings. It goes to colleagues here at IMC who master the new/emergent skills of making a Web Forum into an effective virtual classroom.

We have a fair number of distance education students who in the early days had a one-on-one relationship with their tutors, literally through the post and by phone. We never headed for video-conferencing because the students could not handle it at their end although some of the new kit coming along will transform that. But with the Web Forums we have got to the stage of insisting that all students must discuss their assignments with one another as well as with their tutors. This means the faculty are developing their coursewares to have trigger questions to be mulled over and their role is now more of that of a moderator than a teacher, so to speak.

The contributions coming from the students have, as you might expect, astounded several of my colleagues. I heard one the other day ask why some of his students took the programme at all if they knew as much as they seemed to from their discussions going along.

Anyway, the secret seems to lie in the trigger questions and then well-timed interventions, even summings up, by the faculty. And, of course, the old fashioned discipline of ye olde class register! Every faculty member is now getting contracted for so many log-ins per week and the students are required to do likewise. And up to 20 per cent of the final assignment marks are being allotted to the ideas and learning from fellow students in the virtual classroom. We have just done it for our 1st Global Doctoral Set.

And, wait for it, the uphill battle to get them all along to the Intramural we expect each year has turned into a demand, almost, from them to tell the faculty what they want to discuss and a nigh 100 per cent turnout at the Intramural without a single chaser. They all simply agreed in the virtual classroom what they wanted and where and when they would be doing it.

Plus ça change :... but I long for *la même chose*; or am I getting old? Actually, I wouldn't have it any other way.

By the way ... I saw the work your Internet R&D man, Noel Wynder, has done on Internet Skills and Communications. Quite excellent. We better watch out here on The Other Side. You'll be flying past us soon if we don't take care. Any more good stuff like that coming along?

Best wishes from Derek and I. Donna".

E-postcard 17: **The quest for business models**
"Donna. We are going to adopt your Master Teacher on the Internet approach, and will be highlighting both best practice and practitioners each year in the future. We are also tackling another issue on which you might have some experience. It's a variation on how can we get Internet visitors to our site to pay for the services we offer, rather than regarding them as complimentary. In fact, we've painted ourselves into a corner that requires some bold lateral thinking to get out. Not the only ones I guess.

During our experimental prototyping phase for our Internet programmes we have sensibly quoted low fees to cover our own learning as well as our clients. Now we have polished our performance, we have got to get a realistic pricing policy in place that covers our ongoing costs and development, as well as recovering the initial investment. In a word, we have got to levy higher fees.

Clients naturally want to keep to the level we collected during prototyping, asserting that they have helped us get to where we are anyway – and I have to agree there is some undeniable truth in that.

Our analysis is that the value we are now adding, not only via the action learning educational process but also via the induction of many client managers to the potential of the Internet *per se*, is well ahead of our competitors. And when you add to this our virtual library/intelligence linkage from ANBAR-EMERALD which you led us to, we are world class. As such, the client has to be brought face-to-face with the actuality of our enduring relationship sooner rather than later.

The mistake we clearly made in the prototype days was not to be strongly up front about the follow-on fees once we had got to smooth functioning. But we

These e-postcards and faxbacks continue forward @http://www. irdc.com/

were so keen to gain the business and to be able to advance the emergent activities that we were unwilling to be that open.

The compromisers here argue that we should be able to build so many more clients in future that we can afford to accommodate our old friends at bargain – even loss-making – fees. But you known me as a marketor to the bone and I don't truly believe we need to do that. It's cowardice. We are going to share the facts and the truths with the clients concerned and get them to help us solve the challenge. There are possibly a number of suggestions they can make that will enable us to play a non-zero sum game together.

The abiding truth here seems to be that the better we deliver the relationship marketing process, the higher the expense incurred. So the discussion of what level of relationship service is optimal for both parties needs to be a mature one, involving sharing of ideas, needs and wants from both sides. In the Gadarene rush to build relationships, it is all too easy to create facile arrangements that cost more than the value they deliver.

As soon as I say this, however, it reminds me of the sort of tricks I so despise among service purveyors who use relationships to engender anxieties with their clients/putative clients and then sell a frequently unending relationship of reassurance in its wake. We've got such nonsense running currently on two fronts I can readily think of – Copyright in the Internet and the year 2000. Both are, of course, substantial issues to be addressed but neither seems to generate much positive thinking. Further, we are all being deliberately alarmed by consultants offering to ward off the evil spirits.

My alternative, therefore, is to eschew any alarmism and focus on the ROI, both for the enterprise itself by using Internet driven action learning management development and for the individual manager's career within and beyond the client. I know we have been weak in this latter area, pandering to a "corporatist approach" to manager development when truly, in the long run, it is each individual's own responsibility.

Any observations will be most welcome – G".

POSTCARDS FROM THE OTHER SIDE

References

Ainslie, M. and Wills, G. (1997) "Designing a quality action learning process for managers", *Journal of Workplace Learning*, Vol. 9 No. 3 (see Chapter 10).

Binns, J. (1996) "Coolsites", @ *http://www. imc. org. uk/imc/news/occpaper/cool. htm*

Butler, J. (1995) "Scholarly resources on the Internet", *Campuswide Information Systems,* Vol. 12 No. 3.

DG XIII/E, (1996) "Strategic development for the European publishing industry towards the Year 2000", EU Luxembourg, e-mail: info2000@echo.lu

Hagel, J. III and Armstrong, A.G. (1997), *Net Gain: Expanding Markets Through Virtual Communities,* Harvard Business School Press, Boston, MA.

Heal, R. (1996), "Virtual conferencing", @*http://www.imc.org.uk/imc/news/occpaper/ internet_conferencing. htm*

Hitchcock, S., Carr, L. and Hall, W. (1996), "A survey of STM on-line journals 1990-1995: the calm before the storm", @*http://journals.ecs.soton.ac.ukzsurvev/survey.html*

Kling, R. and Covey, L. (1995) "Electronic journals and legitimate media in the systems of scholarly communication", *Information Society*, Vol. 11 No. 4, pp. 261-71.

Marshall, C.C., Shipman, F.M. and McCall, R.J. (1995), "Making largescale information resources serve communities of practice", *Journal of Information Management*, Vol. 11 No. 4, pp. 65-86.

Mathieu, R.G. and Woodward, R.L. (1995), "Data integrity on the Internet: implications for management", *Information Management and Computer Security*, Vol. 3 No. 2.

Morris, S. (1995), "The future of journal publishing", *Interlending and Document Supply,* Vol. 23 No. 4.

Mumford, A. (1997), *Action Learning at Work*, Gower Press, Aldershot.

Peters, J. (1996), "Quality management as a brand building strategy", *Training for Quality*, Vol. 11 No. 1.

Reeves, T. (1995), "Rogue learning on the company reservation", *@]http:/Twww.imc.org. uk/imc/news/occpaper/reeves.htm*

Wills, G. (1995), "Embracing electronic publishing", *The Learning Organization*, Vol. 2 No. 4.

Wills, G. (1997) "Re-engineering knowledge logistics", *Logistics Information Management*, Vol. 10 No. 3 (see Chapter 3).

Wills, G. and Wiles, C. (1996), "Author authority in the ascendancy", *Innovation and Learning in Education*, Vol. 2 No. 2, pp. 18-25 (see Chapter 5).

Wills. M. and Wills, G. (1996), "The ins and outs of electronic publishing", *Journal of Business and Industrial Marketing*, Vol. 11 No. 1 (see Chapter 4).

Illustrative Hyperlinks on the Internet

ANBAR Coolsites @http://www.anbar.co.uk/coolsite/

ANBAR Electronic Intelligence @http://www.anbar.co.uk/anbar. htm

Asia Pacific Management Forum @http://www. mcb.co.uk/apmforum/nethome.htm

Convenors' Forum @http://www. mcb.co.uk/forumconvenors/

Design & Process Monographs @http://www.imc.org.uk/imc/news/occpaper/occpaper. htm

Human Resource Network @http://www. mcb.co.uk/hrn/nethome.htm

International Management Centres @http://www.imc.org.uk/imc/home.htm

Internet Free Press @http://www.free-press.com/free-press.htm

Internet Research Forum @http://www.mcb.co.uk/internet/

Internet Research and Development Centre @http://www.imc.org.ul/irdc/

IQMA @http://www.openhouse.org.uk/iqma/

ISO 9002 for IMC @http://www.imc.org.uk/imc/charters/iso9002.htm

Library Link @http://www.mcb.co.uk/liblink/nethome.htm

Literati Club @http://www.mcb.co.uk/literati/nethome.htm

Marketing & Logistics Global Forum @http://www.mcb.co.uk/mlf/

On-line Bookshop @http://www.amazon.com

Practitioner Doctorate/DBA @http://www.imc.org.uk/services/coursewa/dba/dbalist.htm

Publisher @http://www.mcb.co.uk/mcbhome.htm

QualNet @http://www.mcb.co.uk/qualnet/nethome.htm

TopMan Forum @http://www.mcb.co.uk/topman/gateway.htm

Virtual Conference Centre @http://www.mcb.co.uk/confhome.htm

These e-postcards
and faxbacks
continue forward
@http://www.
irdc.com/

Chapter 2

Embracing electronic publishing

The grandest revolution in the capture and dissemination of emerging academic and professional knowledge and information since Caxton developed his printing press is now virtually on us. The challenges addressed by medieval scribes as they saw their customized documents replaced by mass produced facsimiles now confront the world's publishers. It is salutary to note that few descendants of those medieval scribes own or control the publishing industry Caxton spawned and which is now to be metamorphosed. From where will the next publishing powers-to-be come?

The revolution is, of course, to be fuelled by the capabilities that electronics will afford over the next decade. It is a revolution that we have all known was coming since the 1970s, but which is only now able to happen. Our abilities to capture, hold and handle the body of knowledge and information, and our ability to transmit and provide user-friendly, domesticated access have now reached their critical threshold. The publishing industries of the world are ready to move from the traditional Caxtonian S-curve to that described as electronic publishing.

As we move, there will be major upheavals throughout the value-added chain of publishing. Authors, editors, typesetters, printers, distribution agencies including the postal services, booksellers, subscription agents, librarians, teachers, researchers and students will all be dramatically affected. The institutions and infrastructures which have been created to realize the Caxtonian industrialization will all be challenged, and many will fall. Almost every one that survives will have to come to terms with substantially changed relativities in the value-adding chain.

There are some who want to believe that the revolution has already happened, because major productivity improvements have already been made using electronic capabilities[1]. But they miss the point. Thus far almost all applications have been devoted to extending the life of the extant institutional infrastructure of Caxtonian publishing. Few of the transformations that electronics makes possible have yet happened. The outstanding editing benefits of word processors have been used to produce good manuscripts more

The ideas and analysis in this article were triggered originally by Bev Bruce and Mathew Wills, with whom I still work, and the late Judith Atherton. Many others have also developed their own electronic publishing initiatives at MCB, most especially Jonathan Barker, Sheridan Brown, Carol Oliver, Timmie Duncan, Deborah Kavanagh, Mike Cross, Ania Kaminska, Chris Wiles and Joanne Hirst. My thanks to them all for fomenting the revolution and hoping they will enjoy its fruits for many years to come.

efficiently. Even when, as now, more and more manuscripts are scanned electronically or disks uploaded and reformatted in the typesetting stage of the value chain, it is frequently to produce masters or film for traditional printed journals.

Such productivity improvements are piecemeal. The behaviour of most academic and professional publishers has been to wait and see what might happen, what someone else might do, rather than to seek to make it happen. At best they have experimented with specific applications. There is a moment in time, however, when such strategies cease to be wise and become potentially disastrous, especially for any relatively small participant in an industry. Indeed, there is much to suggest that smallness can be turned to major advantage at such moments of technological change. Smallness can permit speed and flexibility of response provided there is leadership, skill and determination to make the transition.

This article is a case study of one such small participant in Caxtonian publishing, as it seeks to anticipate the metamorphosis that electronic publishing will surely bring – soon. It describes a determination to move from being a reactive, piecemeal user of electronics to enhance traditional productivities to a proactive, leadership role in the new electronics' institutionalization for publishing.

Anticipatory learning
The enterprise concerned, MCB University Press, which publishes this journal and 130 more besides, is a learning organization of some considerable age and continuity. We have more than a few achievements behind us[2,3]. Yet we have not attempted thus far to anticipate anything so profoundly new as the metamorphosis of the entire industry we work within. It would seem to be the supreme test!

It has been our policy, and it has been effective, to plan to act over no further a horizon than two to three years. Now we anticipate that we must look ten years ahead to discern what might be the new institutionalizations and where we might make our new contribution. And if that is to be so, we should attempt now to ensure that we are in the optimal places, at the right time, with the right products and services, or nearly so.

"Nearly so" seems to us the best we could or should hope for from such an exercise in anticipatory learning. To believe we could do better and arrive at a single forecast, then seek to organize all enterprise to achieve it, is dangerous thinking in a period of structural discontinuities.

Our determination to anticipate is therefore driven not by a belief that we can get it right, but rather by an awareness that we cannot afford not to flex. We are also deeply aware – from the literature we publish in our journals and the conferences we attend – that nobody is yet certain what the significant questions are for the future of electronic publishing, let alone what the answers will eventually turn out to be. Those who acknowledge they know least seem to be adrift, like us, in three broad areas of publishing:

(1) What might be the formulation of the mainstream knowledge and information electronic products and services themselves – will it be articles *per se*, collections/issues or keyworded abstracts leading to full publication and interactive contact?

(2) What will be the channels of capture, transformation and dissemination through which the products and services will flow, that is, the future of publishing logistics?

(3) What will be the transformed framework and ownership of the incoming supply chain of knowledge and information from, and in interaction with, authors?

As an enterprise, we are also adrift in the field of financial reconstruction which, although not the subject of this case study, provides a vitally important reason for anticipatory learning. Our financial reconstruction is designed to give the rising generation of managers and staff the opportunity to participate in the equity of the company, with a view to a Stock Market listing, in no more than five years' time. Any projections of the performance of our enterprise over that timescale must take a view of our ability to face up to and benefit from electronic publishing. The future growth of the enterprise cannot be reliably based on any extrapolation of its achievements over the past 28 years. Nor can future rates of return on investment be taken for granted. Our success thus far can give us the confidence to believe we can make the transformation only if we are willing to identify the significant questions and attempt to action learn our way to satisfactory answers.

We resolved, as suggested by Gutman[4], to begin by developing scenarios of possible future institutionalization of the electronics publishing industry. In doing so, we have combined the glimpses and dreams of others with a careful review of analogous situations in other industries[5].

What is electronic publishing?

It is surely paradoxical that we can, even at this stage, be asking such a question. But it needs a clear and simple answer. For our purposes, we let it mean:

> The exploitation of electronics in any and every cost-effective and cost-beneficial way that can facilitate the processes of publishing.

And we then let publishing, for our purposes, mean:

> Conceiving, creating, capturing, transforming, disseminating, archiving, searching and retrieving academic and professional knowledge and information.

Stated thus, all the piecemeal changes which have already been mentioned, such as authors' word-processed manuscripts, and their scanning or disk uploading into typesetters' own PCs in turn, are a vital part of electronic publishing. And so they are. But they are not the revolution. They are not the metamorphosis.

To gain a glimpse of what is meant by revolution, of the structural changes that are coming, marketers typically point to the changing roles in our society of horse, canal, railway, ocean liner, tramp steamer, container ship and airplane. They point to the infrastructures each brought, of stabling and posthouses and inns, of docks, stations and airports. And they ask what those infrastructures do or are today. We can equally well point to the Corn Exchanges we see in many cities today and ponder what function they formerly filled, and who does that now and where. Do planes land on docks, do containers go from the same ports as tramp steamers did?

Closer to our publishing homeland we can reflect on the revolution that has overtaken the theatre, music hall and cinema, and on their new relationship to today's television channels and video player-driven selectivity. Beyond what we have just lived through, there are the readily anticipated futures from satellite technologies and interactive terminals built on fibre-optic cabling. Yesterday's theatres, music halls and cinemas are mainly bingo halls, supermarkets, car tyre and exhaust replacement centres, or they are gone. Many of the few that have survived anxiously seek to preserve ancient values and approaches, with the same chances of success as King Canute had in his day. Very few have found their renaissant role in the new realities: what they have found is that "both going out and being entertained live" are two enduring values, but they must be repositioned and remerchandized to prosper.

A metamorphosed Hollywood has survived, with some of its great production companies. Cartoon-driven enterprises have taken a surprising lead in unexpected ways through video games and jurassic-era films. The A and B film archives have been resurrected with new value and significance for the hungry new media, too. But the distribution channels for reaching viewers are as likely to be satellite TV or the new street-corner video-hire shop as the four-studio, new-style cinema. Many businessmen watch more films while travelling by aeroplane than at any other time.

What fate, then, awaits the library we know and love, that awesome warehouse for the books and journals we publish, or for its professional librarians? Can it become the intellectual's video/CD corner shop of the future and, if so, should it be seen as a socialized service as our Victorian forebears believed it should for the Caxtonian era? And where are the "canal narrowboat No. 1s" in the new scheme of things?

What fate awaits face-to-face teachers, their students and researchers and authors who use today's libraries on traditional high cost university campuses? Already, in many countries, the advent and growth of open or TV-based universities has challenged much of the institutional infrastructure. Knowledge and information can and are delivered at an electronic distance through wholly new channels. Britain's Open University is already the largest and most cost-effective tertiary institution in Britain today, with an international outreach no others can attempt, let alone match. From 1996, when all its students must be modem-linked to the university, they add the ability for interactive, if not real-time, education to their offering; this in addition to their 20-year alliances for book publishing and with BBC TV.

What fate awaits the publishing houses and their editors and staffs, their agents and distributors? How can we discern the future but "through a glass, darkly"?

Total systems and multimedia perspectives are required

We can begin to comprehend what the revolution of electronic publishing might be only when we take a total systems approach to the process of publishing. It must take us outside the artefacts and institutions we traditionally have, to see what they do for us and why they have been cost-effective and cost-beneficial for Caxton's era. We must not allow either the legacy or the institutional inertia of Caxton's era to blind us to the real opportunities for electronics to be used in the service of publishing.

As we step outside the traditional artefacts, it quickly becomes clear that there are several other ways in which what we have called publishing is achieved, and that these, too, can and will be part of the new electronic publishing *Gestalt*. The most obvious is face-to-face education. To this can be added professional conferring and workshop discussions. Such acts of publishing (normally called teaching and communication), can be consummated either physically or electronically – and in combination with ever-changing relativities.

The TV set in our own homes has become perhaps the most powerful means of publishing in the world today. The BBC is already far more than a broadcaster. It publishes books, videos, the *Radio Times*, has computer-manufacturing joint ventures and text data services. TV at large has already taken up the major, formalized role in adult and continuing education, let alone its contribution through myriad documentaries, music and travel programmes and news coverage. We can but contemplate what were the sources of such information for our predecessors, before Baird and his followers created and domesticated TV. Taylor[6] has observed that the domestication of TV worldwide is the template of what is just about to come for electronic publishing at large. Half a century ago, although it had been prototyped, TV was accessed only by very few. Today, virtually the whole of the world's population, in excess of four billion, can see and hear and learn what goes on across the world or via space probes just a few seconds after it happens.

Faxed messages and e-mail cross the globe at a similar breakneck speed, achieving what postal services and photocopying never could. Telephone and satellite systems transmit knowledge and information at the press of a button. Internet structures enable us to access electronic databases across the world on an interactive basis.

All these are publishing artefacts. Academic and professional publishing in the future cannot and will not be defined in terms of the traditional artefacts we have which Caxton called publishing. A new *Gestalt* is coming. The future will see a realignment of the relativities of audiovisual media. The great contemporary institutions of publishing – the educational system, broadcasting, professional conferring and printed publications – will be transformed. They cannot sustain their quasi-discrete relativities in the age of electronic publishing.

The new *Gestalt* will arise from a total publishing systems' perspective, and from a total or multimedia electronics' perspective. This latter point is already beginning to be exemplified in encyclopaedia publishing. It is showing how knowledge archive publishing can be accomplished via multimedia CD-ROMs, using audio and moving pictures that draw substantially on animated cartoon techniques. Online connections for continuous updates are already being developed.

Seeking and finding the new multimedia publishing artefacts requires the same patience and imagination and skill that radio producers brought to their profession, then cinema and TV producers, too. Together the optimal, cost-beneficial, electronic publishing effect will be achieved. To those who point to failures or cul-de-sacs thus far, and seek to ridicule the attempts, the warning must be to indulge yourself only at your peril. The relative failure, for instance, of simultaneous text conferencing thus far, and of video conferencing, simply makes all the more likely the development of virtual reality conferencing in the future. What will be cost-beneficial and good for such conferencing will surely also be good for the multimedia presentation of academic and professional findings; spectacularly so for case studies and experimental results. And when the opportunity for orderly questioning and supplementary debate is added, Caxtonian separation between a printed article and a conference or workshop discussion will become pointless and ineffective.

The obvious fact that the impacts of electronic publishing will not happen on a particular day of the week or month of the year, but will arrive piece by piece as a jigsaw comes together, does nothing to lessen its metamorphosing role or to allow us time to wait and see.

Envisaging electronic publishing scenarios

Realizing that the electronic publishing transformation will be systemic, not merely a matter of greater Caxtonian productivities, and that electronic publishing ushers in multimedia capabilities, are the vital precursors to addressing our anticipatory learning challenge.

In that realization we have proceeded with our first attempt to envisage scenarios over relevant commercial timescales. These in turn enable us to determine flexible strategies for multimedia product and service development, to determine which channels to participate in with creative enthusiasm, and to explore how to evolve our author relationships – the three fields, it will be recalled, where we currently feel most adrift.

Scenario 1: the authors' co-operative

One of the most compelling scenarios for the future of electronic publishing is best characterized as Marxist. It envisages the world's academics and professionals in free-flow communication with one another. Publishers, editors and libraries have all withered away[7].

At its simplest expression, any author anywhere may enter a contribution into a catalogued database, and all others who wish to access it can readily search on the basis of keywords. For effective current awareness, and for any useful and efficient search to be made, however, it is readily apparent that some form of cataloguing will be required and that maintained and evolved over time by some agency – which is really what a traditional paper publisher is doing when journals are created and tended. Several examples have emerged in the USA already and in the UK at the University of Newcastle.

The traditional analogue is perhaps most closely provided by professional associations with their own non-profit publishing divisions or by the university department with a strong research interest. But most observers do not anticipate the authors' co-operative as the dominant mode of electronic academic and professional publishing, although a renaissance within the professions may well occur. Commercial publishers have grown and flourished to hold a major market share because they have brought risk finance, professional publishing skills, continuity, determination and critical mass to the process. These are likely to be enduring needs in any metamorphosed publishing systems. Many professional associations and university departments have in recent years either sold over their publishing activities to, or joint ventured with, commercial publishers.

What such a scenario does, above all else, is to remind all publishers that the author is indeed the creative talent triggering the flow of knowledge and information. Editors, referees and publishers can structure and embellish their works, but authors are the origin.

Conclusion. All publishers should deeply honour their authors' contribution. They must work to identify with the needs, aspirations and, indeed, the self-actualization of authors. The "best authors" must be found and served through whatever media may be most appropriate. The best authors must be partners with the publishers as they make the transition to electronic publishing. The publisher who loses touch with authors and who fails to create and sustain the very best relationships with them is at great risk.

Scenario 2: consolidator/retailer/agent oligopsony
The great majority of academic and professional journals are consolidated in the distribution by institutions known as "library agents". Four majors dominate the world in their distribution channels, namely EBSCO, Dawsons-Faxons, Swets and Zeitlinger and Blackwells. Two are of British origin, one Dutch and the other from the USA, but all have built global businesses. They consolidate both the ordering and the delivery processes for libraries. This gives them a very strong place in the value-added chain of knowledge and information because they are in touch both with the suppliers of journals and with the libraries which stock them and make them available to readers and researchers.

Their position in the channel of distribution gives them the opportunity to move into retailing, and they have done so. They are now increasingly seeking to provide document delivery services to libraries and to act as CD-ROM service centres. They are also exploring how to extend that role as a retailing node in knowledge and information networks for high frequency EDI rather than for CD-ROM. If they can pursue this route successfully, the same channels of distribution can be used effectively, i.e. their strong relationships with libraries[8].

The downside of this consolidation/ retailing role is that the institution of a library on a widespread basis would disappear. Librarians should not be expected to vote for this scenario, particularly as all staff formerly served by a local library will be able to use Internet to gain convenient access via the library agents.

The alternative manifestation of this scenario is that agents themselves disappear and libraries enter into direct access with publishers, either individually or through the consortia that have been emerging in recent years.

Of these two manifestations, the former – the triumph of the agents – seems more likely because of their economies of operation and sheer breadth of coverage. This will be reinforced by the cost pressures on universities as attendances grow and the inevitable breakdown of traditional, campus-driven university teaching. It would also greatly suit the convenience of publishers, provided an appropriate balance of power can be established and maintained.

In today's parlance, the advent of networked desktop computers means the supply of knowledge and information can be contracted out. The emergence of a desktop library, provided it is driven by intelligent software, can be expected to receive a heartfelt welcome from authors and subscribers alike.

Conclusion. With the exception of the archival/museum function, and an advisory role on how to use the available knowledge and information services, this scenario has no secure place for libraries as we know them. They face the same fate as overwhelmed Corn, Wool and Stock Exchanges before them.

Library agents have the size and the determination to metamorphose into major electronic nodes in the value-added channel and are already doing so. Without the librarian's presence it will be critically important to discern how publisher-agent's relationships will change. Agents, who have traditionally been the servants of librarians, will be unable to achieve the same level of relationship with millions of users worldwide of their consolidated services. They can be expected to squeeze publishers on pricing and service level issues and to dictate common standards of presentation and formatting to ease access by end users.

This scenario allows for the emergence of new publishers and for the handling of products from the authors' co-operative scenario without difficulty.

Scenario 3: publisher oligopoly
There is already considerable evidence that the academic and professional publishing world is moving further towards domination by a few large houses.

In other publishing media, such as TV and newspapers, this has long given cause for concern to politicians, for example, in countries as far apart as Australia and Italy, albeit the trend is largely a result of the underlying economics of information gathering, advertising reach and distribution channels.

It should not be anticipated that oligopolistic ownership will constitute a threat to the freedom and discretion of information gathering processes; but it could certainly have major impacts on access to distribution channels and to advertising revenues on which more than a few professional journals depend. Small or new initiative publishers would probably need to find sponsors from among the oligopolists, and as soon as they were successful could expect to be acquired.

Conclusion. It seems reasonable to expect the trend to oligopoly to continue, but that there will not be any substantial pressure to integrate forward at the expense of library agents or of libraries themselves. It is likely to continue to be in the best interest of the end user to be able to access knowledge and information from a single supplier, in other words a consolidator/retailer, rather than from several or myriad suppliers.

The countervailing power of oligopoly suppliers to match the oligopsony of the library agents will be beneficial for suppliers unless and until agents develop their own sources of knowledge and information (cf. grocery retailers' own label brands). This has already appeared, and more is likely to come in the field of searching/document delivery services. All suppliers of primary knowledge and information, but particularly suppliers of secondary services, will need to adopt high value-added approaches to maintain, let alone build, their positions when such markets emerge.

Scenario 4: entrants from other directions
The history of great industries faced with such systemic metamorphosis is that an outside player, with none of the traditional assumptions or intellectual baggage, enters and dominates – blending acquisition of the crucial industry assets with externally derived advantages. The obvious major candidates already on the scene seem to be:

- Commercial conference organizers who extend their static-dynamic phenomena of face-to-face conferring, via intelligent database strategies, to become both ongoing conference and virtual reality providers. Such continuous publishing of knowledge and information is an electronic extension of their currently fragmented pre-conference materials and post-conference proceedings or papers. Such a networking framework can readily include quality control systems and would make the dissemination of academic and professional knowledge and information to professionals considerably more convenient and user friendly. Such an entrant has no need of libraries, agents or publishers – authors/speakers are the asset to be tapped.

- Software providers who have continuing relations and systems compatibility with the PCs in use within the academic and professional communities across the world. Microsoft is an obvious example, but less obvious are those closely involved in developing computer-aided design systems. All are involved with the animation of text and data. Such entrants would have no need of the traditional value-added chain, and already have substantial credibility with authors and users of their software. They are already wiser and more knowledgeable about the implementation and use of systems than are those involved in any of the other scenarios.

- Video and TV educational providers who can readily extend their activities to encompass academic and professional knowledge and information at its frontiers for current awareness. This is perceived as a logical outgrowth of the use of multimedia electronic publishing for books, particularly texts and games. Those with a lifetime experience of video cartoons are strongly placed.

- Fringe providers who are at the periphery of traditional publishing, but have a strong asset that is empathetic with the new electronic publishing[9]. University Microfilms has long, grey literature and archival searching traditions, and is skilled at providing knowledge and information on a secondary basis, including document delivery. The British Library is a major document delivery house. Such approaches have already led to a challenge to subscription publishing of journals in favour of retrieval of single articles identified by current awareness searches in myriad ways.

Conclusion. Any or all of these entrants pose a major threat and would undoubtedly proceed through acquisition or joint venturing with existing traditional participants in the publishing process. There are also likely to be other entrants not yet envisaged.

They can be expected to be totally cavalier towards sacred cows, such as subscription services and library institutions, as they proceed. Whenever they threaten traditional ways of doing business, they must be treated with utmost respect. The acid test is not whether their approach makes traditional sense but whether the new entrant can make viable, cost-effective inroads. Whatever their message might be, whatever added value they offer, must be embraced.

Crafting our embrace
Our enterprise is not, as already noted, a major volume player on the traditional global publishing scene. We have only 130 journals from a world total in excess of 50,000. But it is at moments such as these that large can be a distinct disadvantage. Provided we strike the right alliances, and invest wisely, our future position can be very different. So how can we craft a future that embraces what electronic publishing has to offer?

The classic yet still enduring starting-point for enterprise must be two questions: what business are we in; and who are our customers? Unless these are answered for the electronic publishing age, 20/20 vision will elude us[10].

We have answered that:

> We are in the business of facilitating the publication of academic and professional knowledge and information in management and engineering by whatever means are most cost beneficial to serve such customers as wish to purchase it at prices that yield a satisfactory profit for our investors.
>
> Our potential customers are those involved in management and engineering who need and want to have that knowledge and information and prefer our manner of publication to any competitive offering available.

Such an answer swiftly leads to consideration of our current, distinctive strengths within such a business. There are many other alternative providers. Why patronize MCB University Press? And then it is necessary to explore what weaknesses we have that must be overcome by development, acquisition or alliance.

Finally, given our analysis above, where do we believe we can and actively want to get in, say, four or five years' time? And how will the diverse scenarios outlined impinge on us as we proceed? How will our perceived strengths and weaknesses affect our success?

We are the scholars and the professionals
The origins of our abiding strength as a publishing house lie in the academies from whence we came as authors and editors. We came into existence to meet a personally felt need in Europe at the time and, over the succeeding 28 years, we have learned how to do and to institutionalize all the other things necessary to have a successful, traditional publishing enterprise. We were at the outset, and strive to remain today, the authors' co-operative. We understand the rationale for academic and professional authorship in the first place; we understand the perceptions and processes of quality assurance; we understand the requirements of current awareness of new knowledge and information; and we understand the processes of searching and researching as the basis for scholarship, authorship and indeed teaching.

Put simply, we know our authors and we know our readers because we have been and still are authors and readers of the services we provide. This intimate intellectual relationship with authors and readers gives the instinctive as well as the analytical understanding needed to create the cost beneficial services of the future. It gives the opportunity to craft the promotional messages that tell our customers worldwide what we are offering. It gives the opportunity to discern what areas of management and engineering, on which we are focused, need and want a new publishing initiative that we can provide.

To sustain, enhance and reinterpret this relationship for electronic publishing is the challenge for management. We must identify and take hold of opportunities, knowing our authors are the supply side of our enterprise and the readers are the demand side, with many of those readers being authors themselves in a virtual publishing circle. It seems that we can and should use electronic publishing as the platform for a seamless relationship, a total knowledge and information flow process from authors as thinking authors to thinking readers. If we can so envision and embrace electronic publishing, we shall provide a very greatly value-enhanced service indeed. And because electronic conferencing allows interactivity and proactivity, we shall achieve what the printed journal medium never could offer alone.

We have the brands
Our second major strength is that we have built and own our own brands, the titles of our journals. There are more than a few words and articles written about materials engineering, about operations management, marketing or managerial psychology, about social economics, cybernetics or library management; but only a modest few are of sufficient and appropriate quality to appear with the endorsement of the journals we own and publish. Inclusion is an accolade, because inclusion is only possible after due process of accreditation, quality assurance, peer review by those whose views are opinions that count, have all been successfully completed.

A newly launched journal cannot carry such a cachet, albeit that its publication by a respected publishing house with a distinguished editor and advisory board can give it a good start in life. It takes many years of high quality contributions by well respected authors, and the support in search of such authors across the world given by senior scholars and reviewers, to create a strong brand that speaks for itself as well as through the words of its authors. Such a brand generates renewals of subscriptions each year which have been the guarantee of courageous publishing in our enterprise for over a quarter of a century.

While accountants are loath to recognize such qualities as assets and count them in an enterprise's balance sheet, that must not blind the enterprise to their value. As we seek to craft our embrace of electronic publishing, we must explore how our brand franchises can be strategically deployed, transferred and strengthened. Any behaviours that weaken or dissipate this intangible strength must be eschewed. Here it is clear, then, that our focus moves on from our 4,000 new authors each year and our archive of 10,000 more before them, to the 130 titles, with their 250 editors and their 1,500 advisers from the academies and professions across the world. We must work with them to develop and extend our brand franchise, to reinterpret its meaning in the era of electronic publishing.

In so far as Scenario 1 can and is carried forward, by authors seeking to work electronically and to build new brands in fields we do or would wish to serve, then we must assist them. We must show them how we can contribute without undermining what they can best do unaided. We must ourselves build the new electronic brands of the next decade, just as we built the strong paper-based brands of the last two decades.

The authors and editors of the next generation of electronic journals will wish to work with us only if we can show we have the capability and competence and an enduring commitment to create and build such new brands. We must answer the simple question posed by the authors' co-operative scenario: what added value does working with a commercial publisher give in building a brand that we cannot do without? We think we know. We think it is of the same order and genre as we have traditionally given to an enthusiastic journal editor when we acquire his or her titles after a few years of independent publication. However, the specifics of the answer must encompass the use of electronic publishing as well.

Davies[10] suggests that we must show how we can turbocharge such publishing, how we can build a better pathway, so the world will want to come to us and to share their ideas and invite us to help them achieve their goals.

We have the technology
Our curiosity in the mid-1970s led us fortuitously to a currently very strong vantage point for our technology base. Our main typesetting and printing house in Yorkshire was reluctant "for union pressures" to make early use of commercial word-processing systems. Accordingly we experimented in-house with several books and, as electronic typesetting became more widely acceptable, we extended our in-house competence. For well over a decade, all typesetting has been accomplished in-house, with printing *per se* bought in. This gave us exactly the right standardization and competence to manage and develop our strategies towards the electronic knowledge and information database our articles represent.

As more and more authors made disks as well as full text versions of their articles available to us, it was apparent that we could transform our electronic typesetting arrangements with Apple Macs and QuarkXpress software. This enabled us both to scan in and reformat authors' contributions but, yet more importantly, to hold all the knowledge and information we had in an article database that could be readily manipulated.

An in-house action learning management development programme[2] identified the need for this transformation, which was accomplished within 12 months – an example of small size working to great advantage. It adopted a structure for our article database that enabled knowledge and information to be tagged for hypertext searching and available for loading to CD-ROM or for online searching, whichever in the fullness of time customers might deem to be most cost beneficial. This constitutes a major commercial advantage for us. We can search and retrieve our entire post-1993 article database on a wholly unified basis, with state-of-the-art approaches available. We can accordingly offer the entire contents as a single "electronic library". Others with contracted-out typesetting services seldom have this unified approach without great additional expense for catching up. Furthermore, our abstracting and keyword services, since 1987, and our annotated bibliographies, back to 1963, make the entire

MCB archive searchable with full text back-up from our own document delivery service.

Finally, we have our own unique thesaurus, applied to both our own and our competitors' titles, for searching the best of the world's literature on management over more than the last 20 years.

We are a global enterprise

The supply of academic and professional knowledge and information in management and engineering is *per se* a global business. Authors for our journals live and work and travel around the world, and the pre-eminent language for them is American English. As such, we are strongly placed with a UK base to access articles and to produce products and services that can be widely read and understood.

These global supply and demand sides to our enterprise also place us well strategically to exploit differential acceptance levels of electronic publishing around the global marketplace. The net benefits achieved by electronic publishing can be anticipated to arrive far sooner in those countries without a strong indigenous state-of-the-art base. Those unable to travel readily or who are geographically at the maximum distance from the location of professional congresses and meetings, and those with limited access to libraries with full coverage of the literature, will find these customary barriers substantially avoided.

The world's major growth markets for knowledge and information in management and engineering are currently to be found in the Far East, together with Russia and the Ukraine. In both, there is little knowledge or information archive available and only limited resources to meet burgeoning demand. The products and services sought will be those which can address their total needs as swiftly as possible.

We have a learning organization

As has been suggested earlier, MCB University Press has one of the strongest and longest-standing learning organization traditions in the UK. There has been continuity of ownership throughout by business school academics and professionals, who brought state-of-the-art educational and development processes to staff at all levels. The company sponsored for 12 years, from 1982, the largest action learning network in the world, International Management Centres[11]. It created the media resources in support of that network, including EDI and BBS systems in the early 1990s, consolidating relationships in more than 30 countries worldwide. It provided a continuous laboratory opportunity in which to observe and share with professional managers in their use/lack of use of management knowledge and information.

These origins and involvements mean that the enterprise has in post, at all senior levels, senior managers who are familiar – if not always comfortable – with anticipatory learning approaches, managers who have been challenged to discern and work through what the future might bring. They have done so within a framework of management that requires them to articulate their own solutions and to take responsibility for making them work, with appropriate accountabilities. These are by no means trivial projects, either. The company's acquisition strategy, its new launches, strategies, the evolution and development of database and relationship marketing, the transformation of the company's technologies, the creation of an Authors' Club – all these have come from staff members and have been turned into reality by those same staff members.

This is the learning organization which now seeks to anticipate electronic publishing. It is this learning organization that has provided and debated the knowledge and information that has gone into this analysis thus far. Furthermore, it is this learning organization that has now begun to design and operationalize our own electronic publishing services.

Yet, to achieve the reconceptualization of the way we satisfy our customers' needs, we are aware that no matter how strong our traditions may be as a learning organization, the worst place to begin to create a new future, and tomorrow's organization, is in the organization we have. We have adopted consistently a prototyping strategy to our electronic initiatives, while at the same time seeking to touch base as far as possible with the managers and staff of the current business patterns.

What we do not have
What we cannot have, even with our determined attempts at prototyping, is the ability to leverage ourselves into the multimedia dimension of electronic publishing without outside help, alliances, joint ventures and projects. As such, we anticipate that our abilities to comprehend what can or might be offered, as well as providing it cost beneficially, requires a great deal more contact[12, 13] with other current and potential participants in the future electronic publishing industries.

From them, we would especially hope to understand the complementarities of the several media – of print, sound, animated and static or dumb images.

From them, also, we must above all seek to understand the social psychology of electronics *per se* and how such knowledge can be applied in the specific fields of our enterprise. How far can electronic conferencing or reviewing go before the processes are so dehumanized as to become counterproductive? How far can the use of dynamic electronics go before the solitude of reading a static/dumb image on a page is preferable? And will such emergent balances remain once discovered, or evolve over time both as electronic capabilities develop and as human experience with electronics grows?

Then, also, we must seek to understand the ergonomics of electronic publishing, the man-machine interface itself. What new or revised concepts are appropriate? How can the whole evolve to be more user friendly?

Finally, what are the services that the emergence of multimedia electronic publishing makes possible that we have not yet realized, or have not been

realizable with traditional media and institutional structures? And how can
they be commercialized?[14].

Sensing the revolution

If times are as significant as we believe they now are, if Caxton is on the verge
of giving way to electronics, then there is only one place to be. We need to be on
the streets with the revolutionaries. We need to sense and feel what is
happening, to share in the dreams of the revolutionaries. And we must do this
both at the supply and the demand ends of the value chain in publishing. And
as we run the streets, we need total systems thinking that alone enables us to
place the events sensed and felt into the electronics perspective.

In this belief, we resolved to learn, as is our tradition, by action learning. We
began straightforwardly enough with an (unsuccessful) floppy disk service. Its
lessons led us directly to a widely acclaimed CD-ROM service for literature
searching across the full range of management literature, known as Anbar
CD-ROM.

In its Mk III electronic version for 1996, it simultaneously extends coverage,
adopts Windows' format and joint ventures with the British Library for full text
document delivery. This latter element represents our first major initiative of
this sort, playing to our own strengths and competence in Anbar from pre-
electronic days in abstracting, thesaurus development and keywording, search
taxonomies and quality assessments. It is an illustration of the alliances
envisaged earlier, with entrants from outside traditional publishing (scenario 4).

Our second major action learning initiative has already been mentioned. It
involved the creation for International Management Centres (IMC) of a global
bulletin-board system (BBS), providing full academic and professional support
to all staff and students. It was a world business school first as a total system
when introduced in July 1994. Furthermore, its knowledge resources provided
to IMC were the first structured application service using Anbar offered by the
enterprise. It exemplified for all just how unique a contribution electronic
publishing could make, by opening up totally new channels of communication
of information on an interactive basis. It also introduced us to the technology of
electronic data interchange (EDI) worldwide with nodal points, and highlighted
the efficiencies and disadvantages of BBS versus Internet as information
highways.

Our third initiative also led inexorably to the exploration of Internet. It
restructured a regular paper-based service to the enterprise's 8,000 library
customers and offered similar benefits on a BBS. The service announced before
publication what the contents of the forthcoming journal issues would be – a
turbocharged service indeed. It rapidly grew to carry the entire portfolio of
MCB journals with editorial details. It also offered an e-mail facility for Anbar
customers, mainly librarians, to order the full text of articles as appropriate.

These three initiatives were prototyped, to the merriment of and with the
customary sideways swipes from, the current managers. Such is to be expected

from the body of knowledge, and to have not been noticed would have been a cause for concern!

The lessons in revolutionary methods were also as expected! We were getting ahead of our customers; and we were making the mistakes associated with being first. We were accordingly faced with the stark choice for our customers and ourselves: did we want to slow down and wait until they were ready and they and we knew what they wanted, or did we wish to stay on the streets with the revolution? And within the enterprise itself, how could we organize so that all would learn what was needed while not distracting them from the current business?

We resolved to create a full-time team led by a well-respected, long serving, senior manager who had successfully tackled technology transfer before, and learned. She had introduced the enterprise to electronic database marketing on her own initiative. It was the team's role to "project manage" strategic initiatives within the extant structure. For their successful implementation the support and the resources of the other managers had to be won. Tough indeed. But the conflict was deliberately designed into the situation so that none could escape the realities of the contribution electronics could make and its power to metamorphose tomorrow's publishing industries. The team, dubbed "electronic publishing initiatives" (EPI), was expected and required to force the pace.

EPI was given two challenges to turn into products and services directly, the confidence to resolve them arising from the three action learning initiatives:

(1) publish creatively all our archives electronically by 1996, initially on CD-ROM; and

(2) exploit the Internet at the leading edge to transform the relationship with subscribers and authors into a continuing dynamic interaction, rather than a series of discrete incidents.

Both were and still are to be accomplished by building the journal brands. The second is to be specifically accomplished through a strengthened relationship with authors and editors.

Both were to be accomplished knowing that we would be ahead of our customers (and staff not within EPI). This implied patience plus educational strategies which, if suitably crafted, could achieve the necessary changes both of organization structure and of attitude with MCB University Press itself[15]. It also implied a determination to be there first rather than to wait and see – the only proviso being that double-loop learning approaches must be applied to all initiatives. Being there first was conceived as the only way to strike important alliances with key players under scenario 4. We also saw it as signalling our determination to work with authors and editors at the leading edge of the revolution – which was exactly where we wished them to perceive us as standing. We resolved to help them to learn, as they helped us to learn, how to make effective use of electronic publishing. In this way, we believed, we could optimize our role in the futures that unfolded.

It's official – the revolution has happened (though we are still working out the details)

True to all revolutions, they have happened before most of us even notice they have begun. What telephones, radio and TV did in their time to shrink the world to a global village has been done with electronic publishing as well.

• The debate now is not about how electronics can and will assist the printed journal but how it will metamorphose the way knowledge and information is:
  - created;
  - accredited;
  - captured;
  - disseminated to give current awareness;
  - stored;
  - searched; and
  - retrieved.

• The debate is about the most cost-beneficial multimedia mix of artefacts for such activities, including electronics of course, but not to the exclusion of other earlier media. What will be new? How will old relativities change?

• The debate is about the metamorphosis of the great knowledge institutions we have known to date. Who will be the new info-mediaries? Whither the universities? Whither the publishing houses? What will be the new cathedrals?

None have any special privileges to remain as they are. MCB University Press will not survive seeking to sell words on paper alone, or electronic impulses for the sake of it. Customers buy benefits and authors will contribute to those multimedia publishers who best achieve their own goals.

In advancing the debate, it seems much clearer now that electronic publishing offers three crucially significant, high value-added opportunities for publishers:

(1) Networking of offices and homes will extend market reach so enormously that there will be a dearth of knowledge and information product and services to offer, just as multichannel TV companies have found. The reach, however, will often be to potential customers unfamiliar with presentation methods of traditional publishers. So success ultimately will go to those who represent most effectively in multimedia format for the widened market and who design imaginative user-focused search procedures.

(2) Authors will want to make use of whatever medium or media mix optimizes their goals of publication to their chosen market. This will

include their need to access the literature by searching and retrieval and to evolve and refine their contribution through peer group processes, whether by conferring or review. Equally, it will include the need for the highest accreditation to be accorded when the work is completed. Success will go to those who achieve these best.

(3) Lazy and unappreciated assets from potential allies in the broad church of publishing can and must be harnessed to survive the revolution and grow strong thereafter. The imagination and lateral thinking to conceptualize what alliances might be and accomplish them are both at a premium as they always are when paradigms shift.

If this be so, if these three elements can be seen to be robust, then they must be the touchstones of MCB University Press corporate strategies and plans for the coming decade. New products and new services must meet their imperatives. Of them all, perhaps the first will be the most troublesome to embrace, because the prospect of the domestication of electronic knowledge and information services is so awesome. It can probably be accomplished only in working with the third.

Can we really anticipate like this?

We are certain the answer is that we can seldom if ever be right first time or for long. We need a strategy for double-loop learning. We need, as a learning organization, to keep our scenarios and our emerging iterative conclusions and the actions they suggest under continuous review[16]. Yet we need to action learn by staying out on the streets, being there, doing it, sensing it all. As such, we will keep asking questions. Furthermore, we can see we need to believe we can survive and re-engineer ourselves well enough to make the financial restructuring of our enterprise, with staff equity participation, an eminently suitable way forward. If our beliefs can be translated into successes then the staff truly will have learned their equity options in the enterprise through the metamorphosis they have accomplished.

Perhaps we can conclude by saying that if we believe we can succeed we probably might, but such a belief does not include any sure anticipation of precisely what our future will be.

References

1. Donovan, B., "Two cheers for the electronic revolution", *The Bookseller*, 21 October 1994, pp. 18-21.
2. Wills, G., "Enabling managerial growth and ownership succession", *Management Decision,* Vol. 30 No. 1, 1992, pp. 10-26.
3. Gore, L., Toledano, K. and Wills, G., "Leading courageous managers on", *Empowerment in Organizations,* Vol. 1 No. 3, 1994, pp. 7-24 (see Chapter 7).
4. Gutman, J.A., "Developing cases and scenarios for anticipatory learning", *Journal of Management Development,* Vol. 12 No. 6, 1993, pp. 53-60.
5. Wind, J., Holland, R. and West, A.P., "Pace setting 21st century enterprises", *7th General Assembly and Exhibition of World Future Society,* 27 June 1993, pp. 1-36.

6. Taylor, J., "The networked home: domestication of information", *RSA Journal,* April 1995, pp. 41-53.

7. Ginsbarg, P., "First steps towards electronic research communication", e-mail address: hep-th@xxx.lanl.gov, 1 August 1994.

8. Duncan, T., Competitors or partners? a study of the changing nature of subscription agents", November 1994, MBA dissertation for International Management Centres, Buckingham, UK.

9. Denkin, D., "IHS succeeds with in-house electronic publishing system", *IMC Journal,* September/October 1991, pp. 13-15.

10. Davies, S., "20 Tips for developing 20/20 vision for businesses", *Journal of Management Development,* Vol. 12 No. 6, 1993, pp. 15-20.

11. Wills, G., "Your enterprise school of management", *Journal of Management Development,* Vol. 12 No. 2, 1993, pp. 3-60.

12. Hall, W. and Davis, H., "Hypermedia link services and their application to multimedia information management", *Information and Software Technology,* Vol. 36 No. 4, 1994, pp. 197-202.

13. Vickers, P. and Martyn, J., "The impact of electronic publishing on library services and resources in the UK", *The British Library LIS Report 102,* 1994.

14. Clement, G.P., "Library without walls", *Internet World,* September 1994.

15. Drew, S.A.W. and Smith, P.A.C., "The learning organisation: change proofing and strategy", *The Learning Organization,* Vol. 2 No. 1, 1995, pp. 4-14.

16. Richardson, B., "Learning contexts and roles for the learning organization", *The Learning Organization,* Vol. 2 No. 1, 1995, pp. 15-33.

Chapter 3

Re-engineering knowledge logistics

The search to understand what electronic academic and professional publishing using the Internet can deliver in terms of cost benefit is well illuminated by the discipline known as logistics. Logistics is the science of the movement or flow of people, materials and information. It had its origins in the military which was until recent times the most concerned with the need to accomplish such major movements in a relatively short space of time. Southworth (1993) and Worsford (1995) exemplify this focus in the context of the Gulf War and the British Army Logistics Corps respectively.

In most human activities, the need for rapid and effective response is not so critically significant and as such merited only limited scientific attention until the last quarter of this century (Kearney, 1991). With the aid of mathematical and statistical approaches linked to computing capability, organizations which expended up to 30 per cent of total cost on achieving effective distribution of their products in the marketplace then began to take the field most seriously.

As a science, it seeks to identify the least cost way of achieving a given level of availability or service to customers. Five key elements have been identified as determining the efficacy of such movements or flows: facilities, unitization, communications, inventory and transportation.

It is the analysis of the trade-offs between each in achieving the requisite availability or service that shows the way to the least cost solution within each cycle (Pohlen and LaLonde, 1994).

This straightforward framework can be well illustrated in commerce:

- Many ladies' fashions in footwear or clothing last for only a short season. Preparations for launch normally have a sufficiently long lead time to enable suppliers to source the items from across the world at the lowest cost. Least cost transportation can be accomplished by sea container. Inventory can be set conservatively in case the fashion does not take off. However, if the fashion catches hold, the supplier needs to know as rapidly as possible both to procure and to deliver more items to retail stores. The additional inventory required can seldom come by least cost transportation or it will arrive too late, nor can it be made far away by a least cost source unless a higher cost transportation mode is used – such as air freight. The speed with which the supplier gets feedback from the information system on sales uptake, enabling him to forecast his need for additional inventory, determines his ability to meet demand. This has seen rocketing growth in marketing/point-of-sale intelligence systems

With *Mathew Wills*,
Logistics Information Management,
Vol. 10 No. 3, 1997

which is apparent any time we shop today with barcode scanners capturing intelligence the moment we buy linked to our customer loyalty card demographics and replenishment cycles (Hagon, 1994).

- Industrial components, whether at manufacture/assembly stage or in the replacement after market, are required in the right place at the right time. However, to achieve that without massive inventory holdings across the marketplace would be impossible without instant access to the information on where a component might be and a willingness to use high cost transportation methods to deliver it. With straightforward statistical analysis of inventory demand, it is normally suitable to hold far less inventory, thereby reducing the risks of obsolescence, loss and damage into the bargain.

The managerial challenge in commerce has been to ensure the consideration of the "total" logistics concept, i.e. view the output of a logistics system as being the result of the trade-off between all five significant elements over an appropriate cycle time (Wenkels, 1995). Their discrete management normally leads to net loss to the enterprise. Simply to minimize transport costs or communications expense will normally give rise to much higher inventory expenditures to achieve a given level of service.

What level of logistics service is requisite?
The notion of a scientific approach seeking to achieve a least cost solution to deliver a requisite level of service or availability begs the question: what level of service is requisite? Not surprisingly, much analysis has addressed this question (Lambert, 1992) – which has an exact equivalent in the currently popular query as to what "quality" a product ought to embody.

The paramount response is, in almost every circumstance, that "reliability/consistency of the level of service offered is more important than speed". If a promise is made, it needs to be 100 per cent certain that it will be kept. The customer for the service can plan accordingly. The precise pattern can be tailored to the customers' needs (Fuller et al., 1993).

The second response, not unnaturally, is that the speed required depends on the context in which the need has arisen. The item designations of A, B and C have been used in inventory control for many years. An A item must be there at all times – oxygen in a hospital or an aeroplane, or engines at the right assembly point in the manufacture of a car. Other items, such as a light bulb in the home, can be regarded as a B or C items. You can borrow one from another room.

Third, the potential substitutability of a given item with an identical/satisfactory competitive alternative will give saliency to the need for a high level of service, indeed preferably 100 per cent availability. It is unlikely that most customers will change their choice of a car, or aeroplane manufacturers their choice of engine because availability is not immediate. However, a chocolate bar or beverage will often be substituted, as will a passenger seat from one airline brand to another.

The traditional logistics model for academic and professional knowledge

In Figure 1, the total logistics flow cycle in traditional publishing is modelled across 12 months.

Academic and professional knowledge first arises in the minds of scholars and practitioners. It is mainly distributed in print or by word of mouth as at conferences, professional workshops or discussions with colleagues. These methods of distribution allow for greater or lesser levels of interaction/proactivity. In a discussion, one can readily proact and evolve a line of analysis and thought; with a book, TV programme or an article, the knowledge is not so readily negotiable. It takes much longer to take up a question and get a response from an author or TV programme contributor than it does to raise a point at a conference or discussion.

Facilities

The retail facilities used to store knowledge are most typically a library or bookstore – although journals will often go directly through the post to readers most particularly in the professions. Publishers use warehouses extensively in the distribution channel and library agents frequently offer consolidation services.

Conference workshop and discussion facilities typically are provided either by hoteliers or by enterprises themselves in countless seminar/meeting rooms around the globe.

Many individuals also create their own libraries, often more an archival stack of previously read or browsed material, in their own homes or offices.

Each of these facilities has its own aura, its own charisma. A library and a bookstore share a quietude not always to be found or even sought at a conference or a workshop, because they are not offering an interactive knowledge consumption process.

Unitization

The unit of knowledge ranges from the well maligned "sound byte" to the impressive tome. In other words, we may seek to acquire access either by simply hearing or seeing a few short sentences, or by necessarily acquiring a book. To gain access via a library, there may be a membership fee and even a fine for taking too long to read/return a particular book.

Journals and professional magazines are frequently sold by annual subscription or at least by the issue. Most conferences expect you to attend the whole at a comprehensive fee or to purchase the printed proceedings afterwards for a far lower fee.

So, while the unit of knowledge is the "byte", its sale and availability are normally in multiple bytes, as a printed volume, an annual subscription to several instalments or an entire conference. A scholarly meeting, to be truly effective, normally requires all its main contributors to be in one place at the same time.

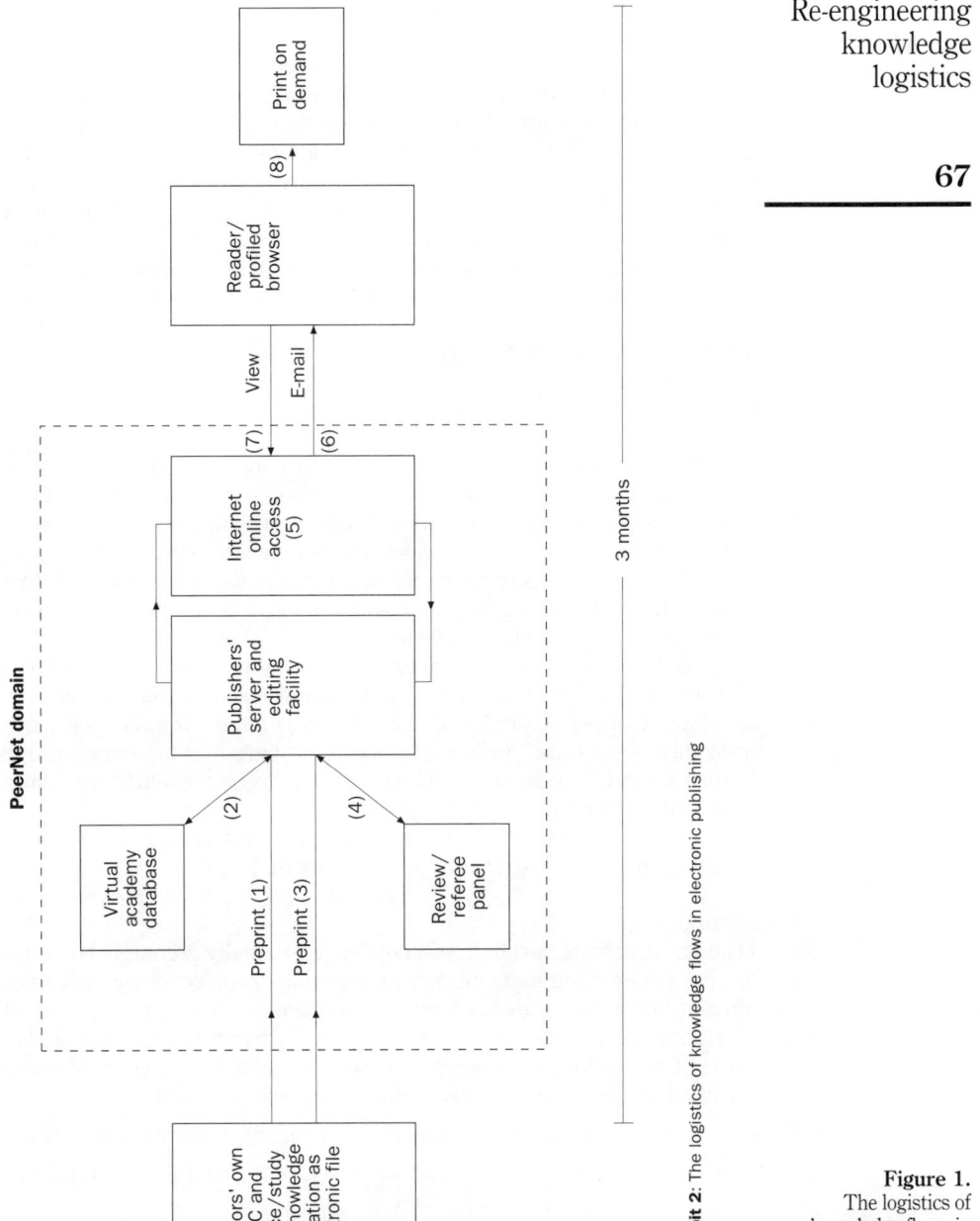

PeerNet domain

Authors' own
PC and
office/study
for knowledge
creation as
electronic file

Virtual
academy
database

Preprint (1)

(2)

Publishers'
server and
editing
facility

Preprint (3)

(4)

Review/
referee
panel

Internet
online
access
(5)

(6)

(7)

View

E-mail

Reader/
profiled
browser

(8)

Print on
demand

3 months

Exhibit 2: The logistics of knowledge flows in electronic publishing

Figure 1.
The logistics of
knowledge flows in
traditional publishing

Communications

In logistics this word is used essentially to describe the awareness available to all potential users of what units of knowledge exist and are available in what facilities or, if they are *en route*, precisely where they are at any given moment. To do this in commerce, large and extremely costly intelligence systems have been developed. Each one of us can ascertain from a bank's autoteller machine what our balances are, and a manufacturer can similarly pinpoint stock locations and levels and where they have reached in transit (Sheombar, 1992).

For knowledge today, we can ascertain the existence of a book-in-print if we know its title/author, of a journal if we know its title, and of an article if we know who wrote it and in what journal. But as to the precise location where we will find it once it has left its publisher, whether it is out on loan or out of stock, and if so how long before we can gain access – such intelligence is not normally readily available.

Inventory

The great majority of libraries will be able to identify whether a book or journal has been acquired and whether it is currently out on loan. So too will many bookshops, albeit on a more reactive basis. Some will know the rate of usage of the knowledge and the average waiting period experienced by customers for availability when it is temporarily out of stock. Knowledge of inventory held elsewhere other than in the particular location will be scant unless a regional information exchange has been established.

The inventory of conferences and workshops on offer is far less well documented. There is no global equivalent of books or journals in print with allocated ISBNs and ISSNs. The inventory is almost totally fragmented and most academics and professionals rely on personal networks and mailed information to access them. They are, however, frequently translated into book/journal format later and then become more accessible.

Finally, the inventory of authors unfinished and/or unsubmitted articles is virtually inaccessible except by word of mouth.

Transportation

The physical transportation of knowledge is usually accomplished ultimately by the reader going to the library or bookshop to collect or by delivery via the post or fax. In the case of conferences and workshops, the transportation is by participants transferring themselves to a faraway, albeit sometimes exotic place. The considerably higher expense involved in the latter mode is regularly justified by the trade-off with access to:

- the most recent knowledge not yet available in books or journals;

- the opportunity to interact and proact and thereby heighten the value of the knowledge transfer; and

- the opportunity to build networks that can of themselves become very efficient media for future learning and knowledge flows.

Requisite service at least cost

The literature of librarianship has frequently reported the application of service levels to the availability of books and journal articles, e.g. by carrying multiple copies or by offering fax or postal delivery services at differential prices. No studies have been located that attempt the total logistics systems analysis with trade-offs costed, mainly because the cost/benefit of academic and professional information and knowledge is grossly underdeveloped. Whereas a commercial product not available can normally lead to substitution of another, this will not usually be the case for knowledge. It is likely to be possible for textbooks but most unlikely for state-of-the-art articles. And the potential value of the missing knowledge must be assessed before any conclusion can be reached as to whether it should in future be provided because it justifies the cost involved.

It is not infrequent to wait for a book from a publisher for a month or more; and to await an article by interlibrary lending for 14-48 days. In such circumstances, it is not surprising that much of the lag time in completing or even preparing a serious draft paper is occasioned by delays in gaining access to particular known items.

This of itself begs the question as to how the potential beneficiary is made aware of its existence, or has a cost-effective way of searching to find all that might be deemed cost beneficial (cp. Vinze, 1991).

It must generally be concluded that users of knowledge are conditioned to accepting that it takes a long time to find something, that the lag time will be inconsistent and that, if it cannot be substituted, the overall output will have to be delayed.

Supply side logistics

I have left until last discussion of the ways in which articles or manuscripts are traditionally procured – whether for books or journals or as conference papers. First, it must be observed that it is an especially lengthy process, rationalized as ensuring high quality but more often than not mainly reflecting inefficiencies. An article or book is normally expected to take 12 months but can take as long as 18 to appear in print. During that time, the actual working time is perhaps four weeks; the remainder is delay or misconnections in the channel of supply.

The author's work in early draft format will be circulated for comments. When finalized, it will go to a reviewer/refereeing panel whose comments will be fed back for incorporation in the manuscript – or, if a rejection, to re-commence the sequence again. Once accepted for publication, the article or book will go through a production process often requiring consolidation into an issue with eight/ten other articles, or await a seasonal catalogue or launch event. These stages, together with repeated proof checking and graphic design contributions, are what take up the 12 months.

The supply side, least cost approach delivering these service levels is, of course, not consistent, not speedy and alas not readily open to substitution of another medium of production. But there is much good news to be heard…

The purpose of this paper is to share just how great the opportunity is to re-engineer both the supply and the demand side of knowledge flows, of knowledge logistics (Ayers, 1995; Badarecco, 1991). In particular, "time to market" as a key contemporary logistics issue is addressed.

Re-engineering supply side knowledge logistics

As has been suggested, supply side flows begin when an author seeks to present research or philosophical outcomes to appropriate academics or professionals. To achieve that task to the necessary quality level inevitably will be an iterative process of formulation, feedback commentary and refinement. It will normally also include a vitally important prior necessity which is to be aware of and in command of what other knowledge exists on the issue concerned – of which more later when we discuss demand side knowledge logistics.

If, however, we commence with the draft article, or preprint, we can explore (see Figure 2) the transformations available from emerging electronic publishing approaches:

Step 1:   The author's preprint is created on a PC which can connect to the Internet.

Step 2:   The author submits to a "virtual academy" of like-minded academics/ professionals for constructive critique and feedback.

Step 3:   Comments are digested and the final version prepared.

Step 4:   The author re-submits for final review and acceptance in a housestyle/ template that conforms to the publisher's specification for Internet publication.

Step 5:   The editor alerts the refereeing panel by e-mail to the article's presence on the journal-protected database and asks for "credit scores" plus comments.

Step 6:   The scores and final comments are fed back to the author for incorporation.

Step 7:   The refereed finalized article or book is published on the Internet.

Step 8:   All interested parties are notified by e-mail of the event and invited to view and/or download.

Clearly links to the earlier critique of traditional knowledge logistics. Accordingly it is not surprising to find such a sequence, known as Project PeerNet, currently under development with MCB University Press for all its journals. It goes further, however, than reducing up to 75 per cent of the lag time in getting articles or books into print. These are some of the major improvements to the very publishing process itself:

 •   The preprints in process are assembled in an electronically available "citation/list/ register" – thereby enabling others to be aware sooner of what

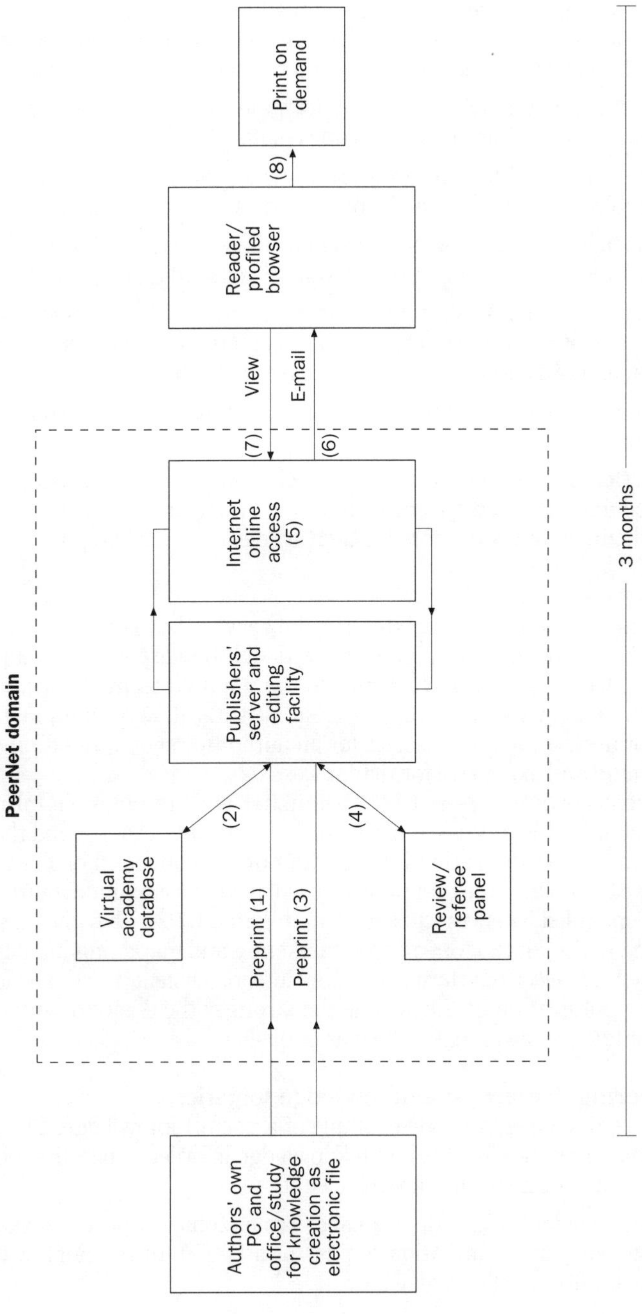

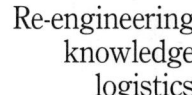

PeerNet domain

Authors' own PC and office/study for knowledge creation as electronic file

Preprint (1)

Preprint (3)

Virtual academy database

(2)

(4)

Review/ referee panel

Publishers' server and editing facility

Internet online access (5)

(7) View

(6) E-mail

Reader/ profiled browser

(8)

Print on demand

3 months

Exhibit 2: The logistics of knowledge flows in electronic publishing

Figure 2.
The logistics of
knowledge flows in
electronic publishing

- The preprints in process are assembled in an electronically available "citation/list/ register" – thereby enabling others to be aware sooner of what is on the way and at the same time to protect the author against plagiarism.

- The online credit scoring in the refereeing process can be expected to produce more consistent and reliable results.

- There is no need to wait for a full issue or a seasonal catalogue to be assembled. Each article can be published as it is accepted.

- The article's abstract can be added immediately to the global abstracting and citation sources rather than awaiting consolidation.

- The author and publisher can proact with interested readers rather than await a visit to a library, a bookshop, a conference or the post to arrive – by e-mail notification to all concerned about publication.

- The interested reader can proact with the author to follow up on a line of argument or analysis.

- The references to the works of others cited in any article or book can be traced and accessed speedily using the newly arriving PII references which are unique indicators for each article.

Verdict
This process will be more consistent in the service levels and the quality judgements experienced. It will be faster. It will be very much cheaper to accomplish once all parties have access to the required technologies. It will, however, almost certainly increase the substitutability of one journal or publisher for another. It will be easier for an author to "shop around" and less time-consuming and expensive for that process too.

The prize accordingly goes to the publisher with the most helpful and supportive supply side process, the best virtual academy resource and the best accreditation of the publication by branding of the collection or list within which it appears. Finally, the best publisher will have the maximum outreach to the author's intended listeners (cp. Brown and Watts, 1992). This gives a whole new meaning to the concept of database marketing and marketing intelligence/ loyalty systems. In electronic terms, the more intended listeners one can alert by e-mail to the publication of the article, the stronger the desire of authors to commit their knowledge to that particular publisher.

Re-engineering demand side knowledge logistics
The author is, of course, just one example of a user of knowledge. The great majority of users are not involved with knowledge in order to benchmark their own nascent contribution. They are there either to:

- take it on board for an educational or professional purpose – a course to follow with an examination to pass or a known professional need to learn or implement a strategy; or

- browse to stay abreast with no specific goal or expectation, which can be satisfied both by finding nothing exceptional and by serendipitiously learning, e.g. database mining approaches (*Computing*, 1992; Grupe and Owrang, 1995).

The ultimate requirements of the author, of the tasked student or professional and of the serendipitious learner will be the full text of the article or book concerned, but all will find it more than helpful to be able to scan/search all the literature available in a designated field speedily and appropriately. The premium here accordingly is on the design of user-friendly, user-conscious search engines of the burgeoning archive of knowledge.

With the benefit of standard general mark-up language (SGML) tagging of electronic text, searching by keywords can readily be accomplished, together with textual mentions. This requires sympathetic thesaurus development within subject fields. Further search criteria can differentiate the recent from the less so, take one to all the contributions of a given author, select by country of origin or reference to industry or corporation, theoretical vs. practical case study as opposed to literature review, and most recently and boldly to the allocation of quality criteria within several key categories.

One can readily see the advantage to being able to select the most recent literature review in a chosen field from a total of 2,000 or more articles on offer. However, even more so, one can readily see the advantage for any professional or educator presenting to an audience to be absolutely up to date with the field concerned so speedily. Before electronic publishing, such a realization was simply impracticable.

The revolution dawning is even greater for the process of updating course materials as in distance learning institutions. On average, the materials they despatch were last updated two-and-a-half years ago. Using automated electronic searching links they can be supplemented with the latest literature as it appears. All that is required is to allocate keywords to each course module and make hyperlinks to the database concerned. This approach is now in operation on a prototype basis. It is no dream; it is a reality and its contribution is awesome.

Similar keyworded or behaviour derived/modelled (Tafti and Nikbatht, 1993) profiling of individual managers can be automated wholly to deliver current awareness. By linking agreed personal key areas of interest to the knowledge database, e-mail alert routines with full text retrieval on demand are being delivered. Complaints that this is tomorrow's junk mail are as well founded, but no more so than the complaint against postal messages in the mailbox today. We are always at liberty to ignore such messages or to have a separate file structure/pending tray for them that does not interfere with the routine of important messaging – just as we may have an ex-directory phone line or a PA/secretary to intercept calls and messages on our behalf.

Knowledge logistics is clearly poised on the threshold of artificial intelligence/expert systems/neural networks for academics and professionals.

The leveraging of the contribution of knowledge workers (Lewis, 1992; Li, 1994; Mockler, 1990; Mykstyn *et al.*, 1994; Osyk and Vijayaraman, 1995; Quinn, 1992, 1993; Ryman-Tubb, 1993; Venugopal and Boets, 1995) has been examined extensively elsewhere in services, manufacturing industry and the professions but not directly for the field of electronic publishing.

Verdict
This process will be both more comprehensive in its ability to trace and retrieve knowledge and will do so almost instantaneously. The data overload which will inevitably arise can be attended to by use of search engines for keywords and abstracts which do not militate against serendipitous browsing or searching but do not necessarily require it.

The benefits of the outreach in distribution first by alerting interested parties in profiled areas but also by immediate availability in the global databases will be of great value to authors both to publicize their work and to elicit interaction.

How key elements in knowledge logistics flows are changing
In Figures 1 and 2, the two logistics flows are modelled. Thirteen discrete phases have been replaced by eight. That presages a major cycle time transformation, but it becomes clear just how wide the re-engineering required truly is when we examine the five key elements of logistics (Persson, 1995).

Facilities
At a minimum, the printing house, the warehouse and the library/bookstore (Schneider, 1994) are either substantially or totally eliminated or meta-morphosed as the move grows apace for electronic publishing. However, no progress can be accomplished until:

- publishers establish the host server for the incoming preprints and their proactive discussion with the virtual academy and referee panel; and
- authors and readers have PC and printer facilities at their disposal for the beginning and end of the process.

It is a classic chicken and egg argument of course. Who moves first? The expense for a publishing house in creating a host server that most of the readers currently do not know how/are not inclined/are unable to use is daunting. But the prizes are paradigmatic if the patient investment can be made. The threatened facilities are slow to respond because they are largely sunk costs and accordingly seem to have only a marginal incremental cost in comparison with a total investment/payback model for the incoming publishing investments. Furthermore, extant budgetary procedures in libraries, for instance, give the funding to the facility and inventory manager not to the reader/browser – to whom knowledge often seems to be a free good – if only you can find it and wait until it arrives. The electronic publishing paradigm requires the reader/browser to have the discretion to buy (within a budget) as appropriate. Institutional inertia and power structures will certainly delay not alleviate change in such circumstances (Smith and Saint-Onge, 1996; Woodside, 1996).

Unitization
The unit of knowledge for the commencement of searching and retrieval in electronic publishing is the keyword not the journal issue or book title. Keywords lead to abstracts and abstracts to full text, with paper wastage eliminated at each stage (Wu and Dunn, 1995). Ultimately, only the articles that cannot be digested on screen will be printed, and only the areas profiled will be drawn to anyone's attention in the first instance if the service offered is proactive.

The need to publish batches of articles for effective traditional publishing and distribution is gone. So is the constraint on how many articles can be published or acquired in any given time period. The remaining challenge which is to quality assure the knowledge offered is not accomplished by a journal *per se* but the quality procedures its editors employ.

Communications
Perhaps the most significant change in the communications element of logistics is the immediate, categorized access to the total body of knowledge for searching and retrieval. This does, of course, include quality guarantees of the knowledge offered. These latter aspects will surely see a transformation from one being driven solely by the editorial review/refereeing procedures to one driven also by the keyword/abstracting services. Many readers or browsers will be unaware of the status of many suppliers of knowledge. The strength of the "retailer branding" via the abstract providers will readily overcome that dilemma (cp. ANBAR Management Intelligence).

The major challenge still to be overcome, with PIIs emerging as the equivalent of barcodes, is how to have a universal referencing system for all units of knowledge for retrieval and within the search engine a uniform thesaurus. These are not unfamiliar problems in the world of knowledge and have been resolved for books (ISBN) and journals/ serials (ISSN). As the appropriate unitization changes, however, the new order has to be addressed. There are, in a range of disciplines such as management/human studies, referencing systems that have been pioneered by the major abstracting services for paper-driven retrieval systems. But these will need to give way to a universally accepted article referencing system and keywording thesaurus.

Once these are in place, the residual role for document supply services, such as are to be found in the British Library and other major archival centres, will surely disappear. Only the archival role will continue to provide for historical research and to act as a backup for lost electronic knowledge. Some electronic publishers are now simply depositing a "golden copy" with these national archives to this end.

This new scenario compares with a helpful librarian who assists a reader/browser to undertake a search beginning with the knowledge in stock in the library concerned, or the bookstore, and then moving to bibliographies and abstracting services one by one – either paper or CD-ROM-based … and then awaiting paper delivery from afar after a time lapse of greater or lesser proportions.

Inventory
The only physical inventory remaining in the electronic publishing paradigm is the printout on demand created by the professed interested party; and the "golden copy" in the national archive and backup for electronic memory.

The startling conclusion for all concerned is that all the warehousing and retailing inventory necessitated by the printed form will eventually be eliminated.

In this context, CD-ROM clearly can be seen as a transitory form of distributed electronic storage/supply because the transportation and connectivities of the total system are not yet in balance. In comparison with only calling on what you need as opposed to the whole inventory on a CD-ROM and a price to match it, it is clear that the future belongs to online access provided that the reader/browser facilities are well able to access it which is closely intertwined with the fifth and final key element in logistics.

Transportation
The use of air and surface postal services to deliver knowledge, increasingly supplemented in the past 30 years first by photocopiers and then by faxes, is destined to be transformed further by server to PC interchange either using the telephone providers or cable networks. Cable provision, linked with television programming, is gaining rapidly on telephone services most especially where telephoning is an expensive access mode. Those centres such as Singapore and Hong Kong with no or almost no charges for local phone calls have shown just how the absence of the cost hurdle can accelerate adoption.

In every scenario presented, the expense of interconnection between PCs and publishers' services using data compression techniques will be well below that of current post or fax services. In many circumstances, dedicated telephone lines and fixed annual access fees are already sufficient to stabilize the transformation cost of knowledge to make it manageable.

The "total" systems challenge
The rapidity with which the re-engineering of the total logistics system of knowledge takes place will be determined by push and pull factors.

Authors will push for more rapid publication with guaranteed quality assurance. Readers and browsers will pull because of the efficiencies and effectiveness of the new approaches, although requiring involvement on their part and competence with the various connectivities.

However, there will be a combination, as has been shown of socio-economic and psycho-technological blockages to the re-engineering process. Where a single unifying entity is able to redesign a substantial part of the total system, using a balanced approach across all fields, progress will be greatest (Kaplan and Norton,

1992; 1996). Any new age logistics system that can build critical mass for the PeerNet construct identified in Figure 2 is in a very strong position to exploit the trade-offs (Wills, 1995; Wills and Wiles, 1996; Wills and Wills, 1996). A number of extant publishers with very strong links both with supply side authors and demand side library and reader markets are now attempting this.

Their six key strategies are to:

(1) put in place the interactive server routines for author preprint manuscript review that attract the best quality knowledge in the first instance;

(2) develop Internet sites that have a critical mass to attract readers and browsers;

(3) build virtual academies-cum-professions as well-classified, benefit-oriented databases – that have their own life in conferencing, listservers and newsgroups;

(4) drive hard for e-mail/Internet connectivities with their end-users via awareness and assistance – cybercafes online and on-site;

(5) create search engines of the highest quality with customer-user-focused search criteria backed by full text delivery online;

(6) empower customers/users to exploit the benefits of the knowledge at their disposal via educational and self-development guidance and support.

A total systems perspective such as this, encompassing strategy (6) above, questions the dividing line between knowledge logistics for publishing and knowledge for education and learning. The new age total knowledge logistics model must expect to see the backward integration of the librarian into knowledge distillation and capture rather than warehousing and stock control – a higher order, higher value-added cognitive task in any event. And it must expect to see the publisher reaching forwards into alliances and joint ventures with educational institutions offering the constant updating of knowledge against given curricula. This can be most clearly achieved in the realm of distance learning where the infrastructure inertia will be least. The high value-added role of the tutor of the future will be to design the curriculum and to facilitate the learning processes in an interpersonal, interactive way. The delivery of mass lectures to students, along with vain attempts to provide adequate levels of tutorial support, will be replaced with tutoring as a caring profession. Staff will be freed up from myriad domestic tasks by the electronic publishing phenomenon.

Those educational enquiries taking place around the globe to see how education and learning can be cost-effectively achieved for the future, as more and more people have higher and higher expectations and needs, have an agenda item they can now address and can take the bold leap forward. They can disaggregate the budget for acquiring information from its central focus and share it among the customers and users at their PCs. They can follow that through by transforming library buildings into tutorial centres. And, finally, they can agree on limited hard copy national archiving strategies, i.e. get rid of the obsolete stockholdings or inventory in educational institutions in favour of downloading on demand. To make progress, however, requires high level managerial skills (Barclay *et al.,* 1994).

The other breakthrough opportunity can be expected among those who have not yet invested in the massive infrastructure of the traditional logistics systems who can go directly to the new paradigms. This will certainly include all major new universities being developed in the Asian region over the next two decades as well as in Latin America and India – which together encompass nearly half the world's population. And it will equally apply to Russia and its own Commonwealth partners as they begin to access the full gamut of Western knowledge.

References

ANBAR Management Intelligence, @http://www.anbar.co.uk/home.htm – which abstracts the world's top 400 management journals monthly and ascribes quality ratings to them on diverse dimensions.

Ayers, J.B. (1995), "What smokestack industries can tell us about re-engineering", *Information Strategy,* Vol. 11 No. 2, Winter.

Badarecco, J.L. (1991), "Alliances speed knowledge transfer", *Planning Review,* Vol. 19 No. 2, March/April.

Barclay, I., Holroyd, P. and Potton, J. (1994), "A sphenomorphic model for the management of innovation in a complex environment", *Leadership and Organization Development Journal,* Vol. 15 No. 7.

Birkett, W. (1995), "Knowledge management", *Chartered Accountants Journal of New Zealand,* Vol. 74 No. 1, February.

Brown, J.H. and Watts, J. (1992), "Enterprise engineering: building 21st century organisations", *Journal of Strategic Information Systems,* Vol. 1 No. 5, December.

Computing (1992), "Nuggets of information: data mining", 22 October.

Fuller, J.B., O'Connor, J. and Rawlinson, R. (1993), "Tailored logistics: the next advantage", *Harvard Business Review,* Vol. 71 No. 3, May/June.

Grupe, F.H. and Owrang, M.M. (1995), "Database mining: discovering new knowledge and competitive advantage", *Information Systems Management,* Vol. 12 No. 4, Autumn.

Hagon, A. (1994), "Electronic trading: the logistics manager's strategic tool", *Logistics Focus,* Vol. 2 No. 5, June.

Kaplan, R.S. and Norton, D.P. (1992), "The balanced scorecard – measures that drive performance", *Harvard Business Review,* Vol. 70 No. 1, January/February.

Kaplan, R.S. and Norton, D.P. (1996), "Using the balanced scorecard as a strategic management system", *Harvard Business Review,* Vol. 74 No. 1, January/February.

Kearney Lecture, (1991), "Reflections on a lifetime in logistics", *Focus on Physical Distribution and Logistics Management,* Vol. 10 No. 7, August/September.

Lambert, D. (1992), "Developing a customer focused logistics strategy", *International Journal of Physical Distribution & Logistics Management,* Vol. 22 No. 6.

Lewis, J. (1992), "Brain gain – artificial intelligence", *Computing,* 18 June.

Li, E.Y. (1994), "Artificial neural networks and their business applications", *Information Management*, Vol. 27 No. 5, November.

Mockler, R.J. (1990), " Non-technical manager's modelling of management decision situations", *Journal of Systems Management,* Vol. 41 No. 5, May.

Mykstyn, P.B., Mykstyn, K. and Raja, M.K. (1994), "Knowledge acquisition skills and traits", *Information Management,* Vol. 26 No. 2, February.

Osyk, B.A. and Vijayaraman, B.S. (1995), "Integrating expert systems and neural networks", *Information Systems Management,* Vol. 12 No. 2, Spring.

Persson, G. (1995), "Logistics process redesign: some useful insights", *International Journal of Logistics Management,* Vol. 6 No. 1.

Pohlen, T.L. and LaLonde, B.J. (1994), " Implementing activity based costing in logistics", *Journal of Business Logistics,* Vol. 15 No. 2.

Quinn, J.B. (1992), "The intelligent enterprise: a new paradigm", *Academy of Management Executive,* Vol. 6 No. 4, November.

Quinn, J.B. (1993), "Managing the intelligent enterprise: knowledge and service based strategies", *Planning Review,* Vol. 21 No. 5, September/October.

Ryman-Tubb, N. (1993), "How Thomas Cook uses neural networks", *Direct Response*, Vol. 12 No. 9, September.

Schneider, F. (1994), "Virtual retailing", *International Trends in Retailing,* Vol. 11 No. 1, July.

Sheombar, H.S. (1992), "EDI-induced redesign of co-ordination in logistics", *International Journal of Physical Distribution & Logistics Management,* Vol. 22 No. 8.

Smith, P.A.C. and Saint-Onge, H. (1996), "The evolutionary organization: avoiding a Titanic fate", *The Learning Organization,* Vol. 3 No. 4.

Southworth, M.S. (1993), "When war wouldn't wait", *Focus on Logistics and Distribution*, Vol. 12 No. 3, April.

Tafti, M.H.A. and Nikbakht, E. (1993), "Neural networks and expert systems", *Information Management and Computer Security,* Vol. 1 No. 1.

Venugopal, V. and Boets, W. (1995), "Intelligent systems for organizational learning", *The Learning Organization,* Vol. 2 No. 3.

Vinze, A.S. (1991), "Performance of a knowledge based system", *Information and Management,* Vol. 21 No. 4, November.

Wenkels, H.-M. (1995), "Re-engineering logistics processes", *European Supply Chain Decisions,* Spring.

Wills, G. (1995), "Embracing electronic publishing", *The Learning Organization*, Vol. 2 No. 4 (see Chapter 2).

Wills, G. and Wiles, C. (1996), "Author authority in the ascendant", *Innovation and Learning in Education,* Vol. 2 No. 1 (see Chapter 5).

Wills, M. and Wills, G. (1996), "The ins and outs of electronic publishing", *Journal of Business and Industrial Marketing,* Vol. 11 No. 1 (see Chapter 4).

Woodside, A.G. (1996), "Theory of rejecting superior, new technologies", *Journal of Business and Industrial Marketing,* Vol. 11 No. 3/4.

Worsford, R. (1995), "Royal Logistics Corps in a changing world", *Logistics Focus*, Vol. 3 No. 4, May.

Wu, H.J. and Dunn, S.C. (1995), "Environmentally responsible logistics systems", *International Journal of Physical Distribution and Logistics Management,* Vol. 52 No. 2.

Chapter 4

The ins and the outs of electronic publishing

Dallas (1995) captures our dilemma neatly when he comments: "Great invention; now what's it for?" The sooner we cease gongoozling the technology of electronic publishing (which still only really works on a good day anyway) the better. Electronic publishing is overdue a healthy dose of world-class marketing thinking and campaigning. What valued benefits in use does it or can it offer to its customers? What new products or services can it offer to deliver those benefits? What prices will be willingly paid? What forms of marketing communication will be most appropriate? How will the products and services be distributed? It can be concluded that, unless these questions are addressed professionally sooner rather than later, a great deal of investment will yield little return – or worse.

Back to basics

So what is electronic publishing? It is readily defined as using computers in the processes of capturing and disseminating knowledge and information. It implies that the parties at each end of the process, and as may be required in the middle, have computers that can converse with one another. The conversations can be either interactive online or static via either disk or fixed time data interchange (EDI).

Such conversations, without computer mediation, are scarcely a new idea. Interactive conversations have always taken place between individuals face to face, and latterly using telephones since the dawn of the twentieth century. Static conversations have taken place through manuscripts since writing began, via printed books since the fifteenth century, through journals, paintings, photographs, cinematography, radio and television.

What is new therefore is the computer's capability to avoid or complement the ways in which we hitherto interacted or in which we held static conversations. The way in which printing avoided and complemented the work of scribes, telephones and radio avoided or complemented the need to travel to hear, the cinema and television avoided or complemented the need to travel to see as well as to hear, so too computers avoid or complement the need…to print to read, to travel to hear or to travel to see. Multimedia publishing using computer mediation encompasses them all, though smell and touch remain absent.

It is salutary to note that all developments thus far have not yet eliminated any antecedent medium – people still like to write, they like to travel to listen, to talk and to see. As such paperless publishing seems as unlikely an outcome as the demise of the cinema, painting, television, radio or the telephone. Yet they will

With Mathew Wills,
Business and Industrial Marketing,
Vol. 1 No. 1, 1997

only live on in a fresh complementarity to what electronic publishing can best provide. Electronic publishing in the fullness of time will force those new complementarities among the earlier technologies, also implying avoidances as well.

Academic and professional publishing

We shall focus in this article on the implications of electronic publishing for a hitherto traditional academic and professional publisher. The challenge we will address is how can and should such an enterprise metamorphose and proceed to market with the products and services electronic publishing now offers.

We shall do so with two customer groups in mind which are characterized here as The Ins and The Outs. The Ins are the creators of new knowledge and information who wish it to be published. We shall call them "authors" throughout – they are the supply-side market. The Outs are those who wish to know and be informed. We shall call them "readers" throughout – they are the demand-side market.

Both customer groups, authors and readers, are already receiving some but certainly not most, of the benefits that electronic publishing can reasonably be expected to offer. Accordingly, in each case we shall model their needs and highlight where electronic publishing offers scope for added value. We shall also be especially concerned with change retardants. These are key elements in the current way of doing things which will inevitably have to be overcome before benefits can be realized. These mainly, but not always, take the form of institutional inertia.

Benefits of electronic publishing to authors

The process of authorship begins with a desire to communicate something the author believes others should readily want to hear. It may well be a philosophical piece or the findings of research conducted. In either case the author will often wish to be able to position the new contribution in the context of what has gone before. (In this respect, the author will be a reader of the already published works of others and the benefits of electronic publishing as discussed extensively below for the reader will apply.)

The traditional model of publication is to circulate a draft among a close group of colleagues whose opinions are valued. Their comments and observations are debated with them and, as appropriate, incorporated in the script. This script is then forwarded to one or more relevant journals for their consideration. Publication in such a journal will ensure recognition within the academy or profession where the author works as well as disseminating the knowledge widely. Each journal will have a procedure for reviewing manuscripts to see whether they meet the declared editorial purposes. In a leading academic journal this will often amount to what is termed "triple blind refereeing" where three reviewers critique the contribution without knowing the author's name. The critique is fed back and again incorporated as appropriate in a revised version of the script (Evans, 1994).

The arrival a decade ago of word processors made the task of incorporating revisions a great deal more efficient than hitherto. It is only recently, however, that any serious attempt has been made to develop a "total" creation/review/ amendment process using the potential of electronic publishing (Harnad, 1990; 1995). As this has been explored it became immediately apparent that electronic publishing can add value at several stages. First, an original draft, often called a preprint, can be greatly assisted by "conferencing" among other interested parties prior to the circulation of the draft. Comments can be gained by e-mail in a fraction of the time paper circulation might take and the constituency for elicited comments can be greatly extended. What is required to accomplish this is a Virtual Academy cum Profession for conferencing and to whom such preprints can be readily circulated.

Onward transmission to the editors of learned journals is also more readily achieved by electronic file transfers, and the editorial refereeing process can be more expeditiously done. Indeed there is no need for an editor to be selective among reviewers. All members of advisory and review boards can be circulated electronically and those who have something to say can self-select. The benefits of such an approach lie not just in its simplification, however. It speeds up the process of review wonderfully so that publication can be achieved much more speedily.

Finally, the publishers of the journal in question are able to have transparent access to the processes, so they are at all times aware of the flow of good articles coming forward for publication.

We are prototyping this process currently at MCB University Press as BetterEd III – our third electronic support system we have made available to our 300-plus editors worldwide. The Virtual Academy cum Profession we have created since 1991, known as Literati Club, is made up of well over 11,000 authors over the past five years who have contributed to our journals together with our editors and their Advisory Boards (Wills and Wiles, 1996). The great majority of incoming contributors have a global body of 20 or more recognized authorities in any focus area whose advice and critique can be sought.

The major change retardant is perceived to be anxiety by editors and authors that such a widespread distribution of preprints may lead to plagiarism. There will surely be occasional instances of this but the benefit of the additional feedback from a body of other interested authors which would not normally be available should more than compensate for such a risk. So too should the speed at which accepted articles can be "officially" published under a journal imprint.

The second retardant, which will in time be overcome, is the lack of e-mail connectivities for and system compatibilities for more than just a few of those needed to be directly involved.

Figure 1 depicts the authoring process which can realistically be completed within no more than three months against a current norm of at least six months. The article is held electronically throughout and simply modified and improved as required. During that shorter timescale a far greater wealth of advice and comment can be expected. It is a win/win outcome.

The ins and the
outs of electronic
publishing

83

Figure 1.
Benefits of electronic
publishing for authors

Publishers are increasingly asked why, if such an integrated electronic publishing process is possible, any publisher is needed at all? Authors after all create the original ideas and most of the reviewing processes are conducted for honoraria rather than full fees.

The answer must surely be that the system itself needs a systems operator (Hitchcock *et al.*, 1996), and no matter who that might be – old style metamorphosed publisher or new gatekeeper – the role must be reliably and competently executed. Furthermore, the publisher is the guardian and developer of journals as brands that guarantee acts of publication as being fully recognized. Authors are but only occasional participants in most such knowledge and information creation systems. They might create one or two articles each year. For maximum credibility most journals seek to be global in their supply-side market search for authors. That requires a continuing search and encouragement process (Wills and Wiles, 1996). Finally, the financial risks of undertaking the whole process have to be underwritten, regardless, and a return will necessarily be expected on the necessary investment.

Benefits of electronic publishing to readers
Readers have two predominant motivations. They want to:

(1) stay up to date, be confident that they are aware of what is developing in their area(s) of interest; and

(2) from time to time they need to search the body of knowledge inside out to find everything worthwhile that has been published on a given topic.

The first motivation is best described as current awareness/browsing and the latter as archival searching (see Figure 2).

Currently, paper printed abstracting journals and printed copies of original journals on library shelves are the main methods used to achieve the goal of current awareness/browsing. Archival searching is achieved by cumulative indices, CD-ROMs or via online searches of major databases (Siddiqui, 1995).

Figure 2.
Benefits of electronic
publishing for readers

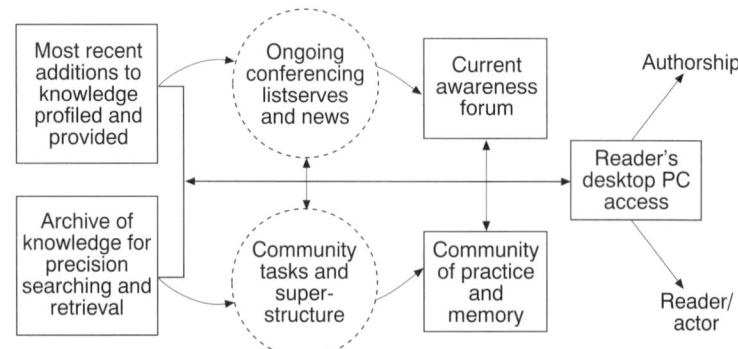

This use of online databases for archival searching is an excellent example of how electronic publishing has already transformed the efficiency of such activities. In medicine, the law, many pure science areas, in news, and increasingly in management, databases are being compiled that speed up the task of locating relevant knowledge and information. When complemented with efficient document delivery, any reader in the world has access to the body of knowledge (Line, 1995).

Current awareness and browsing demand less than the total archive, of course. They demand the most recent. Furthermore, because the process is one of skimming rather than reading, they require an easily and speedily digested format. Abstracts and précis approaches are most commonly used.

In both cases, because the body of knowledge is potentially so extensive, and expanding all the while, selective search approaches are required in which the reader can have complete confidence. If a contribution has been included can we be sure it is good? If it has been omitted can we be sure it is not significant? In such circumstances the branding and authority of the online database or abstracting service for current awareness have to be of the highest repute (Sandelands, 1995). Instant gratification with just some of the body of knowledge will seldom be sufficient.

Perhaps the most significant benefit of all that seems to be emerging for readers via electronic publishing is the ability to enlarge the canvas of both current awareness/browsing and archival searching. Because of the appetite of computers for data, and their ability to structure and manipulate them, two new paradigms have emerged.

Electronic current awareness and browsing services can encompass the widest range of communications – not just journal articles but also newspaper and magazine pieces, radio/TV items, new products and services, professional conferences/colloquia and so forth (Towell and Towell, 1995). Forums and Listserves are emerging making specific use of the Internet that can reasonably be expected to be worth a weekly or fortnightly visit by any academic or professional involved in the field.

Within the archival area, the parallel development for readers is what Marshall *et al.* (1995) have described as "communities of practice". Community memories for electronic communities can be created that formalize what is durable from the forum processes and also capture what has hitherto been classified as the "grey" literature, i.e. information not captured in the mainstream media but nonetheless well worthy of capture.

The concept of community memory goes further than simply expanding the archive, however. It provides an interpretative superstructure at any moment in time of how a given electronic community looks at information and intelligence. In effect it captures the cultural context and gives incremental normalization to the body of knowledge – particularly driven by its significant tasks. Our first pioneering Internet/journal-sponsored conferences led by Davies (1996) and Peters (1996) achieved exactly this in the fields of electronic peer review and telecommuting.

Both these reader supportive, beneficial constructs, the forum for current awareness and browsing and the communities of practice, are at variance with the unitization of knowledge and information entering the electronic publishing system from authors. Authors create one article at a time, but the appeal of the forum and the communities of practice is that they have a quantum of their own that attracts. They each provide crucial bridges between large-scale digital libraries and the day-by-day workshops of professions and communities.

To simplify the paradox, authors require accreditation of their article input per se. Readers require accreditation of their chosen current awareness service and archive at large so they can be reassured they are not missing what they should have encountered.

The final major contribution electronic publishing offers the reader is access anytime, anywhere and printing on demand. The need for a visit to the library, the inconvenience of waiting a month for full text retrieval, or for access to a CD drive – all these are potentially things of the past. Each and every reader can have networked desktop, even at home, access to the world's very finest archives and current awareness/browsing services. Each and every reader can have profiled current awareness and personal agent searching for the information and intelligence that illuminates their salient requirements.

So what are the change retardants? The major retardant is the institutional infrastructure already in place of librarians, traditional publishers and information gatekeepers. Not only are their roles under threat of almost total metamorphosis but also the real estate represented by the library buildings and the stocks and archives can no longer have the priorities of the past 100 years (Line, 1995). Budgetary patterns of allocation require drastic realignment which all takes a great deal of leadership and time. Each desktop needs to be networked. Each academic and professional requires training to take advantage of the extensive benefits of electronic publishing.

The good news is that the budgets increasingly misallocated to the old models of information dissemination for readers are sufficient to fund the electronic publishing paradigm. Electronic publishing promises much greater value for money than traditional media. The value added by electronic publishing will be quite remarkable as the widened extent of the market among readers makes them much better informed and drastically reduces the average unit cost of furnishing the information and intelligence to the reader.

Campus-wide, company-wide, country-wide and global networks are now arriving increasingly across the world. As buildings were once wired for electricity then phones, increasingly we are being wired for intelligent operation.

Managing the transition

As a traditional publisher with a going concern customer base and institutional structure, we have to ask ourselves how we can market our way through the stark business discontinuity we face (Wills, 1995). And while doing so, we must be watchful for those who are not seeking to make such a transition but are starting either with a clean slate or from a quite different entry point to our market. Software providers already have desktop presence but not too much content. Libraries and their agents have major budgets that can be differently deployed.

One element on which we have been able to reach a deliberate decision is the way we capture all knowledge bytes as a publisher. Since 1993 all bytes are in SGML ready for all media of publishing – print, disk or online.

Second, having gained the advantage of a common byte structure in our knowledge database we are comfortable to provide all three current major media of dissemination for our customers in libraries and for our readers as ultimate consumers. So we now offer all major journals we publish in each medium. This has required that new services be added to our range – Cybercafes, (c/f Hilton Hotels reported in Lynch, 1995), Technical Help Desks and Network Licences have been three of the most obviously vital.

But the unique benefits added by computer mediation have also meant we have driven hard for the development of search engines to take maximum advantages of the knowledge contained on our CD-ROMs or available online via Internet. We have also backtracked into our archive to 1989 to make it electronically searchable via keywords and abstracts with full text retrieval guaranteed. For so long as the copyright laws retard the availability of other publishers full text online, we have also forged an alliance with the British Library to ensure world-class full text service back-up from its Lending Division (de Bellaigue, 1995; Morris, 1995). This approach will also be indispensable for supplying references cited by authors that readers wish to locate.

Online access to the knowledge bytes we publish has one further and significant implication. While CD-ROM and printed journals can perforce only be dispatched from time to time, and EDI realized only on a batched basis, with our Internet World Wide Web strategy we are able to offer continuous, instant publishing of any and all accredited articles. Not only therefore have authors reduced the delays to getting their contributions accepted but also once accepted they can be immediately distributed to all readers – with e-mail notification of online arrival on a routine basis to all paid subscribers and as appropriate on a listserver basis.

For many readers online access to published knowledge resources will be more expensive incrementally than utilizing either a CD-ROM or a printed journal. If the lapse in time since publication online is not that significant, it will for many be several more years before online use of networked services becomes the norm. For the duration, accordingly, our product and service strategy must remain to be all things for all media – comprehensively.

Yet at the same time, just as we are developing our competence in the electronic publishing media, we are also convinced we must discern the new complementary role (rather than alternative role) for paper. What is the unique contribution of paper-based knowledge to readers who already have electronic access and have been notified by e-mail when additions to their subscription on a continuous publication basis have been made? E-mail overload is, of course, one concern to be expected but solutions on sorting and sifting e-mails will certainly arrive via computer personal agents if not by traditional PAs or secretaries.

Put another way, if each day we were e-mailed by our personal computer agent, against our profiled interests and favourites, a schedule for watching TV, would we still want to have a copy of the *TV Guide* and what would it say to us?

To summarize then, an ideal transitional scenario from a publisher perspective for 1997 could well be:

- continuous publishing of articles on the Internet (Butler, 1995) in a Forum context, with e-mail notification to subscribers as each article is posted;
- CD-ROM produced every four or six months;
- printed journals circulated quarterly with comprehensive pagination.

Such an outcome maximizes the encouragement and support of online Internet publishing futures, and makes the best uses of electronic features. It nevertheless wholly protects the paper only/paper residual role and CD-ROM preferences for the journal's subscribers. It minimizes the multiple media production expense by working from common knowledge bytes held in SGML and supported by Adobe Acrobat.

Any scenario which wholly sustains paper exactly as it was prior to the advent of electronic publishing is counter-productive. The electronic media will be accompanied by new style paper complements, designed specifically to enhance the value gained from them.

All such transitional arrangements can, of course, proceed regardless of any substantial progress among authors to reap their major benefits from electronic publishing. The supply-side market can only gain the full benefits when virtually all are able to participate from their own offices. Emphasis on the upside gains of speed of reviewing cycle and speed of publication would seem to be the appropriate approach to stimulate such a result. At the end of the day, however, as a publisher we must ensure we garner in the best knowledge bytes available regardless in our focus areas of how an author feels most comfortable when submitting them.

Eschewing a product-driven mindset

Cross and Smith (1995) "cringe every time (they) see an (enterprise) transpose a product-driven mindset on to this new, most customer-focused of communications media". They were referring to the commercial gold-rush on to the Internet not by publishers but by countless companies taking a home page or two. But their criticism applies to most publishers too.

If we stop at ensuring the latest articles are published continuously, and e-mail notification is routinely sent to subscribers, we are missing out on most of the current awareness and browsing potential of electronic publishing not to mention the supply-side market.

As such, with over 150 journals predominantly in management we have resolved to champion forums focused not on our journal products *per se* but on the broad frame of current awareness within which the latest articles in any one of them falls. This has already led us to create Hospitality Forum, QualNet, Human Resource Network, Librarians' Link, our authors' online Literati Club, Futures Forum, Asia Pacific Management Forum, Top Managers' Forum and a "Showcase Electronic Journal" for 84 of our journals – known as our EMERALD suite @ http://www.mcb.co.uk/ emrld.htm – and there will be many more to follow.

Each of these is managed by an individual who is typically not an editor of one of our journals. It avowedly gathers together a cluster of our titles as a key element, but encompasses kindred spirits in an Open House/Cool Site/Yellow Pages format. Each of these forums is intended to become a worthy and significant "First Stop for Managers" in the given focus area.

We certainly have not achieved what we want yet. We have only just begun. But the logic of current awareness across a field of interest drives our initiative which the possession of our own Internet server has readily facilitated.

Our "First Stop" strategy is based on the well researched conclusion that most academics and managers have many other, better things to do with their working hours than search or surf the Internet. We recognize that in the fullness of time our First Stop strategy can/will be displaced by personal agents searching and sifting through cyberspace for ideas and topics of interest against preordained/even fuzzy search criteria. However, we believe there will be an enduring interest in a "Hitch-hiker's Guide" that has personalized endorsement – just as Egon Ronay or Michelin may guide and reassure us in our selections of restaurants or hotels. Knowbots or robosearchers do indeed have a major role to play not to the exclusion of a human intervention contribution but to supplement its efficacy and relieve its monotony. They will open up opportunities that would otherwise not be comprehended (Willson, 1995).

Accordingly, if such is to be the core strategy to meet current awareness needs, what is the ideal forum design? What features will make it irresistible to go there, enjoyable to be there, and most conducive to regular return visits? The geographers among us immediately identified the challenges addressed by central place theory in urban studies, of retail gravitation analyses, social physics, and shopping centre development. The conceptualization and knowledge derived in those areas are both readily applicable. These can in turn be complemented by the understanding of in-store dynamics that retailers have developed. The resultant list of desiderata has begun to emerge as follows:

- Critical mass is vital. Goldsmiths will prosper best by being in Gold Street. Quality management providers will prosper best by being on Quality Street. Competitors prosper by working cheek by jowl – for themselves and for secondary and/or complementary providers which support their enterprise (Weston, 1995).

- Readers/searchers for current awareness will gravitate towards that location within convenient reach that promises the greatest potential attractiveness in the field concerned. The Internet vitiates all concept of physical distance leaving only "subjective" perceptions of proximity to analyse – which will be affected *inter alia* by elegance, value in use characteristics and quantum.

- The First Stop needs to be the place to go, and the place to be. It is the place other folk who matter in the field are already. It is the talked about spot. It is the socially accepted thing to do – and to become this it needs appropriate endorsements.

- The First Stop also needs to be positioned in and among other germane and browsable/eavesdropping activities. While the visitor may not want to use the others they supplement the feel good factor when visiting the focused destination.

- The assortment and the selections of current awareness provided will need careful and imaginative planning as all retailers know. The site must have a very high in-stock percentage of what visitors know they want to know. Plus, a good record for surprisingly intriguing yet relevant awareness too. It will have an atmosphere, culture, and service convenience to go with it that make any visitor satisfied with their visit and more than willing to return.

- Finally, although the intention to re-visit may well be created, there is no substitute for what retailers call the "principle of interception" – a deterministic future encounter with readers because of the way they are going about their lives quite outwith a formal decision to First Stop at our site. The newspaper kiosk at the railway station, and impulse items at a supermarket checkout demonstrate a powerful use of physical interception to sell items that would otherwise often be overlooked or forgotten.

In cyberspace, interception already takes the form of hyperlinking from one site to another and from one page to another – designed into the cyberstructure of current awareness rather than left to bookmarked chance. Interception now increasingly uses listserves. Add to these the opportunity to use e-mail judiciously not to sell but to trailer current awareness items to known visitors or for prospecting campaigns, and the analogy with retailing is almost complete.

A few, but only a few, examples have emerged thus far of genuine attempts to create First Stops as opposed to hanging the information archive. Cross and Smith (1995) report an excellent prototype at PrePRESS Main Street, launched in April 1995. The site is an information centre for its industry.

PrePRESS Solutions sells image-setting systems, software services and training to printers, graphic designers/artists and desktop publishers. Main Street has the feel of a friendly small town community. It is helpful and rich with free resources and advice. And, surely, it also has goods and services for sale as well:

- Main Street Gazette is updated daily with industry news – and it hyperlinks to all other relevant newspapers on the Internet too (see Edmonston, 1995);

- Café Noir is the place to go for newsgroups, chats and links to relevant professional services;

- The Convention Centre presents full reports from the grey literature and from trade fairs and shows worldwide;

- Classified Ads is an exchange and mart, gratis, for small advertisers of equipment for sale;

- PrePRESS DIRECT is the company's catalogue – with four languages available – for ordering and browsing;

- The Print Shop is the technical help desk with FAQs and addresses and the chance to leave enquiries by e-mail;

- The Art Gallery shows some of the impressive work done by PrePRESS customers in graphic design and layouts;

- The Reference Library is replete with downloadable guidance on PrePRESS products.

Robert Trenkamp, the President of PrePRESS Solutions, declares he wants to sell more by using the Internet as a powerful marketing device. He has created a gathering place/First Stop which he believes will expand awareness, interaction and credibility for PrePRESS. As such, he sees it primarily as an integrated marketing activity. Furthermore, he places great emphasis on the customer participation it affords. Not only are customers encouraged to hang their work in The Art Gallery, they are also encouraged to ask advice from The Print Shop and use the Classified Ads service too.

Access is increasing every month. Trenkamp is tracking the traffic patterns on his site as good retail store managers do. "A simple restructure of part of the site can produce instant change in the number of people attracted to a given function. In due course we will become the site which truly helps the customer to buy rather than overtly seeks to sell" (see Oh, 1995; Troiana, 1995).

The coherence this all provides as an extension of the database marketing revolution of the 1980s is breathtaking (Cross, 1994). It offers the chance to calibrate and interact with as well as to profile the customers we serve. (Bruce *et al.*, 1994; Rushbrook, 1995; Wills *et al.*, 1991).

Why the hard sell is not working on the Internet

PrePRESS has exemplified what can be achieved with an integrated marketing strategy using the Internet. Such reports are, however, still very much the exception. The academic literature on marketing on the Internet is virtually non-existent and professional trade literature offers either grandstanding hype (Levinson and Rubin, 1995) or tales of modest results. Thompson and Kaul (1995) suggest a global sales fetish is driving the Internet.

The modest results repay careful scrutiny to begin to discern what behaviours strike a chord. They have typically but not solely arisen in fields already acceptive of mail order and distant delivery. Computer shareware has sold well (Hoge, 1994; Smith and Linton, 1995). Sporting trophies (Andelman, 1995), flowers and peanut products (Dieckmann, 1995) have been successful. So too have books in general, computer books in particular (Crawford, 1994), rare books, music CDs (Fried-Cassorla, 1995) and antiques. Airline and English Channel Tunnel Eurostar bookings have found a place.

Successes have also been achieved in the way pioneered by Digital to launch and most particularly to demonstrate their new Alpha chip and the workstations around it (Andelman, 1995). Over $15 million new sales follow-through is credited to his approach over 12 months from launch – these orders did not come online but were from salesforce follow-up, using Internet in a complementary, integrated way including customer service advice as well.

Kimberly Clark in New Zealand (Kaye, 1995) has made similar use on a demonstration basis with its salesforce.

Foresight Promotions (Oliver and Smith, 1995) have integrated film clips on Internet with voucher requests to attend local screenings – gaining 65,000 hits in three months and meeting their screening targets. On a further occasion they held talk-ins by Internet Relay Chat with a film producer and writer.

Guinness and McDonald's (Dwek, 1995) have both used Internet humorously to build their image with their target markets.

Retail electronic kiosks are being increasingly used as a wholly focused example of Cybercafés. They normally complement other traditional marketing communications approaches (Partridge, 1995). One of the most effective thus far offers a multimedia virtual reality holiday before bookings are made, thereby giving customers great reassurance about a major annual decision. Argos, the UK catalog retailer, is providing kiosks for customers to browse the stock items to supplement their printed catalogues. Other retailers, as with Kimberly Clark in New Zealand, train their sales staffs to walk their customers through on screen. Federal Express gives customers direct access to monitor the movement of their shipments to customers or between centres.

From Canada, Living City (Crawford, 1994) has been introduced to provide a graphical rendering of an imaginary city. You simply walk around the major business districts and chose where to linger.

Online comparison shopping for well established brands has succeeded on a sale or return basis. Malls like Barclay Square (Jolly, 1995) and the much longer established in the USA such as QVC are confident they have a powerful new medium for the conduct of e-commerce as it is increasingly called.

Batty and Lee (1995) describe ways in which, nonetheless, e-commerce can be expected to reach way beyond traditional mail order paradigms and their catalogues. They adjudge current offerings as providing low differentiation and comparabilities for shoppers. They argue that e-commerce at its best will combine the satisfaction of shopping around with the objectivity of comparative testing approaches. By offering prospects the opportunity to identify and weigh their own choice criteria without an oppressive "best buy" categorization, interactivity will be optimized within a global storehouse. If it's all too much trouble, the "recommended" option will always be there.

Another harbinger of e-commerce comes from an Irish success story. Iona Technologies (Hanley, 1994) gains 75 percent of its sales over the Internet. Incorporated and staffed in Dublin, it exports 99 per cent of its services as a computer software company. Its major market is the USA where it gains nearly three-quarters of its revenues yet has only one staff member in an office in San Francisco.

This inventory of reports in the literature of the modestly good news all points, as observed at the start of this section, to the PrePRESS philosophy. Computer mediation in marketing is not an end in itself; it is one more, and uniquely different, element of the marketing mix. Use of the Internet in marketing the products and services of any enterprise will only succeed when a careful evaluation of how it adds value to the customer concerned has been conducted and acted on. It can only be deployed within the total marketing framework – and most particularly the interactive communications mix.

Almost every example reported emphasizes complementary use, whether to aid and supplement promotions or field sales staffs, or to transform customer services as with Federal Express. What is crystal clear is that computer mediation in marketing will involve most considerable dis-intermediation of many current actors in market channels whether they are generating transactions or achieving deliveries (Weiner and Brown, 1995). As new patterns evolve unitization decisions will change and pricing practices with them (Rowley, 1995). New, secure payment systems will emerge to support transactions finalized online (Higgins, 1995).

Our "revolutionary" agenda

Accordingly, while a great deal has yet to metamorphose (Wills, 1995), it seems wholly unlikely that our academic and professional publishing house will find any way forward using an Internet hard sell. We must stay on the streets with the electronic revolutionaries, so we can identify and run with those novel, value added characteristics of electronic publishing that customers will learn to expect and will consequently be willing to invest in (Hitchcock et al., 1996). These are, perhaps surprisingly to some after 15 years or more of uncertainty, now becoming reasonably clear and we are incorporating many of them into the heart of our marketing strategies already:

- Authors – are being offered the benefits of computer mediation to ensure we gain access to the best contributions available including preprint discussions and conferencing NewsGroups and Listserves.
- Readers – are being offered the fullest range of availabilities on paper, CD-ROM and online from commonly held knowledge bytes with imaginative search engines.
- Customer services – are being facilitated and greatly improved on the Internet including replies to frequently asked questions or FAQs, advanced notice of articles and e-mail notification on publication.
- Promotional campaigns – are being supplemented and complemented by access to our Internet site with commonly held messages, sample/ showcase articles and ordering services.
- Database marketing – is being linked for interactive customer development to improve feedback to editors and authors on readers' satisfaction with and guidance for the future.
- Current awareness and browsing – are being constructed as a discrete service on an alliance/joint venture basis with all parties who can together create a potent First Stop in the areas where we are deeply involved in publishing.

The novel value-adding dimensions of electronic publishing where we are laggards, where we are underaware, where we must seek to find commercially viable ways to develop, seem to include the following:

- Greater opportunities for interactivity – both giving showcases for readers and authors' own successes and for feedback and talkback between authors, editors and readers, supplementing our NewsGroup conferencing services (Cooper and Kleinschmidt, 1995).
- Audio visual digitized publication – we have been unable thus far to discern how to create virtual reality for readers in our publishing patterns but we must (Hitchcock *et al.*, 1996).
- Integrated communications strategies – that consciously plan to link rather than complement and parallel. This area must include promotional communications and paper/electronic knowledge and information dissemination as well as hyperlinking to cited references within articles, thereby raising the burden of search from the reader.
- Application of the principles of retailing – to site traffic and merchandising (Lohtia *et al.*, 1995; Smith, 1995).

It is and must continue to be our goal to transform our traditional publishing house into a contentedly, wholly embracing "electronic" publishing enterprise. Much has been achieved but the revolution continues – exhilarating yet fraught.

The new institutionalization is still some way off but its dimensions are becoming apparent.

References and further reading

Andelman, D.A. (1995), "Betting on the 'Net", *Sales and Marketing Management,* June, pp. 47-59.

Batty, J.B. II and Lee, R.M. (1995), "InterShop: enhancing the vendor/customer dialectic in electronic shopping", *Journal of Management Information Systems*, Vol. 11 No. 4, Spring, pp. 9-31.

Brown, R. (1993), "Write right first time", accessible via Literati Club on http://www.mcb.co.uk/literati/articles.htm

Bruce, B., Jordan, T. and Wills, G. (1994), "Realising the benefits of a marketing intelligentsia", *Marketing Intelligence & Planning,* Vol. 12 No. 6, pp. 21-34.

Butler, J. (1995), "Scholarly resources on the Internet", *Campus-wide Information Systems*, Vol. 12 No. 3, pp. 12-14.

Cooper, R. and Kleinschmidt, E.J. (1995), "Performance typologies of new product projects", *Industrial Marketing Management,* Vol. 24, pp. 439-56.

Crawford, M. (1994), "Around the world on a shoestring", *Canadian Business*, December, pp. 83-91.

Cross, R. (1994), "Internet: the missing marketing medium found", *Direct Marketing,* October, pp. 20-3.

Cross, R. and Smith, J. (1995), "Internet marketing that works for customers", *Direct Marketing,* August, pp. 22-3, 51.

Dallas, K. (1995), "Great invention; now what's it for?", *Marketing,* June 29, pp. 27-8.

Davies, R. (1996), *Career Development International,* forthcoming, also accessible on line at http://www.mcb.co.uk/literati/articles.htm.

de Bellaigue, E. (1995), "Copyright and content: a contemporary conundrum", *The Bookseller,* September 1, pp. 18-21.

Dieckmann, M. (1995), "Doing business on the world-wide web", *Managing Office Technology,* June, pp. 41-3.

Duranceau, E.F. (Ed.) (1995), "The economics of electronic publishing", *Serials Review,* Vol. 21 No. 1, pp. 77-90.

Dwek, R. (1995), "Spiders on the web", *Marketing,* August 17, p. 29.

Edmonston, J. (1995), "Rapid changes alter media planning", *Business Marketing*, August, p. 18.

Evans, P. (1994), "The peer review process", accessible via Literati Club on http://www.mcb.co.uk/literati/articles.htm

Fried-Cassorla, A. (1995), "Successful marketing on the Internet: a user's guide", *Direct Marketing,* Pt 1, February, pp. 23-6; Pt 2, March, pp. 39-42.

Hanley, C. (1994), "Casting the Net", *Computing,* 10 November, pp. 30-31.

Harnad, S. (1990), "Scholarly skywriting and the prepublication continuum in scientific enquiry", *Psychological Science,* Vol. 1, pp. 342-3.

Harnad, S. (1995), "Premature polls", VPIEJ-L Usenet NewsGroup archive, 14 August, see http://burg.lib.vt.edu/ejournals/vpiej-l/vpiej-l9508.html

Higgins, K.T. (1995), "Into the Cyberspace", *Credit Card Management,* February, pp. 35-42.

Hitchcock, S., Carr, L. and Hall, W. (1996), "A survey of STM online journals 1990-1995: the calm before the storm", accessible at http://journas.ecs.soton.ac.uk/survey/ survey.html

Hoge, C. Snr (1994), "Winning in electronic marketing", *Direct Marketing,* May, pp. 27-9.

Jolly, D. (1995), "Under your thumb", *Direct Response,* August, pp. 31-7.

Kaye, M. (1995), "Marketing in the new millennium", *Marketing,* March, pp. 14-20.

Levinson, J.C. and Rubin, C. (1995), *Guerrilla Marketing on the Internet,* Piatkus, London.

Line, M.B. (1995), "Access as a substitute for holdings: false ideal, costly reality", *OCLC Systems & Services,* Vol. 11 No. 4, pp. 11-13.

Lohtia, R., Johnston, W.J. and Aab, L. (1995), "Business to business advertising", *Industrial Marketing Management,* Vol. 24, pp. 369-78.

Lynch, M. (1995), "Top business comes to grips with Internet challenge", *Computing,* July, Vol. 20, pp. 16-17.

Marshall, C.C., Shipman, F.M. III and McCall, R.J. (1995), "Making large-scale information resources serve communities of practice", *Journal of Management Information Systems,* Vol. 11 No. 4, Spring, pp. 65-86.

Morris, S. (1995), "The future of journal publishing", *Interlending & Document Supply,* Vol. 23 No. 4, pp. 20-2.

Oh, D. (1995), "The role of prototyping in bottom-up approaches to systems development: the ideal", *Campus-wide Information Systems,* Vol. 12 No. 4, pp. 12-18.

Oliver, R. and Smith, T. (1995), "Reaching generation X", *Direct Response,* May, pp. 33-40.

Partridge, C. (1995) "Talking shop", *Computing,* 2 February, pp. 26-7.

Peters, J. (1996), "The Hundred Years' War has started", *Management Decision,* Vol. 34 No. 1, pp. 54-59, also accessible at http://www.mcb.co.uk/literati/articles.htm

Rowley, J. (1995), "Issues in pricing strategies for electronic information", *Pricing Strategy & Practice,* Vol. 3 No. 2, pp. 4-13.

Rushbrook, L. (1995), "Buying in the Cybermarket", *Marketing,* March 2, pp. 20-21.

Sandelands, E. (1995), "Which journal? The politics of where to publish", accessible via Literati Club on URL http://www.mcb.co.uk/literati/articles.htm

Siddiqui, M.A. (1995), "A study of the effect of CD indexes on online searching in a science and engineering library", *OCLC Systems & Services,* Vol. 11 No. 4, pp. 14-21.

Smith, G.E. (1995), "Framing product design: using design communications to facilitate user learning", *Journal of Business & Industrial Marketing,* Vol. 10 No. 5, pp. 6-21.

Smith, J. and Linton, F. (1995), "Electronic commerce = no more keiretsu?", *Computing Japan,* No. 14, pp. 15-20.

Thompson, D.G. and Kaul, A. (1995), "The corporate response to the World-wide Web… towards a hypermarketing phenomenon", June, pp. 1-20, accessible at http://www. studassoc.utas.edu. au/webmaster/wwwmarketing

Towell, J.F. and Towell, E.R. (1995), "Internet conferencing with networked virtual environments", *Internet Research,* Vol. 5 No. 3, pp. 15-22.

Troiano, M. (1995), "Surfing in the USA", *Campaign,* 7 July, p. 30.

Weiner, E. and Brown, A. (1995), "The new marketplace", *The Futurist,* May/June, pp. 12-22.

Weston, R. (1995), "Five ways to do business on the Internet", *Inc Technology,* No. 3, pp. 75-7.

Wills, G. (1995) , "Embracing electronic publishing", *The Learning Organization,* Vol. 2 No. 4, pp. 14-26.

Wills, G. and Wiles, C. (1996), "Author authority in the ascendant", *Innovation and Learning in Education,* Vol. 2, also accessible at http://www.mcb.co.uk/literati/articles.htm

Wills, G., Bruce, B. and Duncan, T. (1991), "Creating a marketing intelligentsia", *Marketing Intelligence & Planning,* Vol. 9 No. 4, pp. 3-20.

Willson, J. (1995), "Enter the Cyberpunk librarian: future directions in Cyberspace", *Library Review,* Vol. 44 No. 8, pp. 63-72.

(Mathew Wills (mwills@mcb.co.uk) is Electronic Publishing Development Manager with MCB University Press at Bradford, UK. Gordon Wills (gordonwills@mcb.co.uk) is Professor of Customer Policy with International Management Centres.)

Chapter 5

Author authority in the ascendant

This article describes the evolution of the world's largest academic and professional author/editor support initiative, known as the Literati Club. It analyses the leading actions and double-loop learning (Argyris, 1993) crafted over five years to achieve a better understanding of authors' needs and publisher value added at the dawn of electronic academic and professional publishing.

We can see now, quite clearly, that the future requires a reconceptualization, not just re-engineering, of academic and professional publishing processes. The supply-side access and demand-side egress boundaries of Caxtonian publishing structures are being progressively removed (Davidow and Malone, 1993). Their removal takes the would-be publisher of the future into direct contact with authors – first, as they put pen to paper (to use an outdated phrase); second as they browse to enhance their subject area authority; and third, as they search, research and draft their own next article. A sellers' market beckons the electronic author who is both supplier and consumer of academic and professional knowledge.

In Caxtonian publishing the supply-side access to print has been stewarded by editorial peer groupings, production editors and capacity constraints. The demand side has been warehoused and accessed under the guidance of a plenitude of professional librarians amid continuing debate on the affordable extent of budgets for books, serials and physical accommodation. The two sides have been mediated by an oligopsony of no more than a dozen library agents worldwide, resolving *forex minutiae,* unit procurement and consolidation of incoming documents. All these participants can only have a future if they metamorphose their roles to add value in electronic publishing.

Unconstrained sharing of knowledge worldwide

The most widely canvassed scenario for academic and professional electronic publishing is what has been dubbed the "Marxist" outcome. All intermediaries will wither away. Authors will post their initial ideas as working papers on the Internet, comments will be proffered by literally any interested party, the author will revise the article after a given time period, and then it joins the globally accessible archive of knowledge having been classified under an agreed protocol for convenient search and retrieval by all interested persons. Knowledge will become a (relatively) free good. The publishers who have profited from the exploitation of authors' copyright will have no value-adding role.

Many Caxtonians dismiss such a notion as fanciful. They believe it is no more likely to succeed than Marxist distribution theory in Soviet societies. But

With Chris Wiles,
Innovation in Education,
Vol. 2 No. 1, 1996

such outright rejection is fraught with danger. There is not a single ingredient in the scenario that is not already in its separate place in the Caxtonian world. And even if the full integrated extent of such a scenario is not realized either in the short or medium term it will very soon begin, before long, to impact on existing organizations.

A supply-side sellers' rather than a buyers' market is potentially in the making. The author in the electronic age will have a much wider range of options available. The erstwhile publisher will have to compete vigorously for the right to distribute and must wholly comprehend the authors' perspective to identify those aspects that they wish to see developed, if the publisher wishes to remain the "preferred" distribution channel.

There is only one realistic response to a sellers' market in any field – the development of greater understanding of how the seller is thinking, and why.

In the beginning: searching for efficiencies

It would be grossly dishonest to claim that it was for this reason that Literati Club was established at MCB University Press in 1991. Its origins were from a staff action learning development programme (Wills, 1992) that concluded an "authors' database" would be helpful for production editors and proof readers. It would afford easy contact with authors as their articles were being typeset.

As soon as the idea was floated, the organization's marketers thought laterally. It could be the way to sell the journal to the library at the author's home institution, if they do not already subscribe. The organization's publishers quickly realized that if the authors were catalogued not simply by name and address and journal, but by article keywords, some readily usable intelligence would be to hand for editors in search of advisory board members, authors for themed issues and for editors to help accomplish new journal launches.

The original inventors in the production area of the company were swept along by the enthusiasm of a host of others who wanted to use it. In the company skunk works the dusty boxes of authors' details from the past three years were opened up, and a "Start up/catch up" authors' database was born, with some 4,000 names. The editors and advisory board members were included too. For the first time the data we had so speedily lost sight of on the key individuals who drove our supply-side market were assembled within a readily analysable structure (Jordan *et al.*, 1994).

Today, nobody knows how we lived without it. The early cynics have long ago become silent. New authors since our initial start up have been added year by year, and some of course deleted. It now holds nearly 10,000 names across the world on whom we have up-to-date information about their research interests, the fields of their written scholarship by keyword retrieval, the details of their supporting information providers, their seniority/volume of publications, their views and opinions on our service to them as authors, their familiarity with and use of the Internet and their attitudes towards up-coming electronic publishing scenarios. Literati Club members have twice rallied to the flag when an entire editorial team walked off the job leaving us stranded, and then started a new competitive journal within months!

All this intelligence made us a lot more efficient; and it certainly needed to because it has been a substantial investment. But we have also become well informed, quite specifically, about what authors and editors want us to do to improve ourselves – both operationally and to enhance the "quality" of the journals we publish.

It was this latter intelligence which has oriented our activities over Literati Club's first five years. We consciously sought to put right the problems and difficulties authors encounter in getting published in journals where they want to appear. The issues when stated are blindingly obvious. They were not raised by great numbers of authors – in all instances by less than 10 per cent ($n =$ 1,377). Our challenge was to seek cost-effective ways of addressing them.

Developing the editors' role
The most significant improvements sought were full square on the editors' desks. We work with leading authorities across the world and they with distinguished advisory boards and reviewers. Frankly, authors opined, some editors were downright inefficient and ineffective. The acknowledgement and review processes could often take 12 months to complete. Furthermore, the criteria by which the articles were being judged were not always clearly articulated or adhered to. The widespread notification of formal editorial plans and, in particular, up-coming themed issues would be of great benefit. Author-meets-editor open discussion meetings at conferences would also be valued so that the above issues could be shared face to face.

Such calls for improvement were not what we might have hoped for. They were not, so to speak, under our direct or internal control. They required us to go outside to evolve and lead the editorial processes of almost volunteer, highly respected individuals scattered around the world. If we had been asked to redesign the page layouts or the outside covers, or the referencing systems – all these we could have done in a few months. But no; they were not the areas for improvement so far as our authors were concerned! They wanted better editors.

We had a history, from June 1986, of publishing briefing notes for editors under the series title *As The Publisher Said to the Editor*. They had first appeared to announce the universal introduction of abstracts and keywords for all journals from 1 January 1987. They had offered advice on a million-and-one ways to generate good articles, particularly in the early years of a journal's life. In 1992, they were pulled together as a 36-page briefing for all editors, new and old alike. Their success since 1986 may well have been the reason why less than 10 per cent of our authors saw fit to suggest improvements – but we wanted to reduce further the 10 per cent.

We now adopted a different approach. Rather than passing on our own publisher-outwards received wisdom we invited the editors themselves to tell us what they got up to. What ways had they developed to ensure that authors got a guaranteed service level on article reviewing turnaround, and how did they ensure the process was fairly conducted against pre-known criteria; as might be expected with hindsight, they amazed us. We received more than a few details of the computer-based databases that individual editors had created to track and automate chase-ups among reviewers. A wealth of guidance notes for reviewers were to be found around our editorial networks. Encouragement to give creative and constructive criticism was enshrined in many of the best.

From this was born Literati Club's own software in 1994, appropriately named BetterEd, which is currently in Mark II, with Mark III only a year away for those able to use Internet electronic peer review processes. The best reviewing procedures were also captured by Evans (1994) and shared throughout the editorial network, and with all authors. We also introduced a notation of "senior" editors among the total body of our editors as our critical reference group for continuing consultative development.

The second major initiative we took to develop our editors' role was to convene annual workshops on the eve of the club's Awards Luncheon held in London each year for each journal's most outstanding author. The briefings and the BetterEd software were complemented by good old-fashioned face-to-face, share and compare, discussion of problems and new ideas. We are never going to underestimate the benefits of such an occasion despite its most considerable expense. Voices from across the world become faces, editors go on to meet the authors whose papers have been deemed best, whom they have seldom encountered before.

Going direct to the authors
Those outstanding authors give us their views on how we can improve and are also themselves exhorted to go home and encourage young authors to take courage and write their first contribution. This we seek to encourage in two important ways. First, we ask all our club's authors to introduce appropriate colleagues to us so we can find the right journal editor to mentor their first contribution. We call this our New Faces campaign which was launched in 1994 and now sees its first cadres in print. It has been widely appreciated, and is a specific example of how MCB University Press is seeking to build and retain its role as a preferred supplier to new authors – a good investment for the longer term.

Second, we have pioneered campus-based authors' workshops worldwide, run by editors and senior MCB University Press staff. They are yet a more constructive helping hand to new faces/authors. Any and all faculty members on significant campuses are welcomed. Guidance entitled, "Write right first time" (Brown, 1993), "As the publisher/editor said to the author" and "How to get into print" and free ranging, open debate are the order of the day. Once again, the appreciation shown is considerable but it is extraordinarily difficult to measure the cost/benefit of it all. We have no doubt other publishers can well benefit from our efforts! The crucially important element accordingly is to ensure that the messages from authors at the workshops are not lost even when no article subsequently arrives for review. In reality, we have learnt that those messages can be more valuable than an additional article or two.

Every workshop has given the same message, whether held in Bournemouth, Cambridge, Florence, Milan, Johannesburg, Sydney, Brisbane, Melbourne, Bangkok, Hong Kong, Singapore, Tokyo, Buenos Aires, Curaçao, São Paulo or Chicago. Face-to-face, worldwide, over 400 authors have now both reinforced the suggestions of the 10 per cent in our 1991 survey and asked us to do all we can to help new authors learn the rules of publishing. They have described how, in most cases, it is well nigh impossible to do justice to the requirements of editors and reviewers because of the lack of cost-effective and timely access.

This latter feedback has spurred us on most considerably as we confront the opportunities afforded by electronic publishing both on CD-ROM and via the Internet. The creation of world-class secondary publishing services and making our own journals available electronically was given top priority in 1994.

And overarching all these messages is the greatest imperative of academic life – publish in the better journals or your career will perish!

Discerning what a quality journal is
To be successful as a publisher, MCB University Press has always realized its journals needed to have prestige and recognition, to have high quality. The challenge we faced was to understand what our authors and our subscribers meant by quality.

There are several league tables of author ratings of journals in our fields, but no objective study of the criteria for judgement or how those judgements accumulate over time. We resolved once again to seek the dimensions of perceived quality from our authors. Using an independent research consortium (Day and Peters, 1994), we were able to identify the 14 most significant elements under three broad descriptors of Prestige, Content and Presentation. Each journal was then rated on the 14 elements to identify their relevant saliency, and then, most importantly, benchmarked against comparator journals to which the author could have sent the article concerned. By a most elegant research design we were enabled to open constructive discussions with our editors worldwide on how they could not only improve their efficiency but also their effectiveness – i.e. their quality.

This latter accomplishment was of equal import for the professional management of MCB University Press as had been the original capture of the Literati Club database. Benchmarking had now emerged as an objective, learnt system. Its arrival empowered our professional managers to give advice to the external, largely other-directed editors of our journals. Indeed, the editors could be encouraged to initiate, analyse and own the evolved benchmarking process themselves. The implications of this across our 150 journals are still being worked through; the realizations are still arising in the busy world of deadlines (Day and Toledano, 1995). Yet it is a vital and urgent agenda, for now and for the future we think we are beginning to perceive in electronic publishing (Wills, 1995).

The maintenance and development of high quality branding by journals seems likely to be critical in electronic publishing, if for no other reason than that capacity constraints of the printed journal can be removed. A single journal brand can aspire to publish "all the good articles" in a field rather than the number that can be contained in six issues a year of traditional style. Or a distributor/secondary publishing brand can aspire to become the arbiter of what is best in the field using its own accreditation processes. *Ad hoc* surveys are increasingly appearing around the world in response to university-level reward structures linked to "quality" publications. Citation indexes exemplify such a trend; ANBAR Electronic Management Intelligence with the British Library is taking it one major step further. It derives its accredited list from the holdings of the major business schools of the world and a panel of distinguished editorial advisers (Sandelands, 1995).

Auditing progress after five years
The internal feel good factor about Literati Club is high; and the external merchandising effect is beyond doubt. But, as has been suggested, it is a major investment, and we needed to plan for the next five years. We have already taken the club onto the Internet as a world first, convened by Robert Brown, to integrate and provide online access to our resources and to our espoused supportive role for authors and editors.

Accordingly, we wanted to evaluate what progress we might have made over our first five years. We revisited our original membership study at start up, altered as few questions as possible, and mailed it in the first six months of 1995. Of club members, 814 responded by the end of August and form the basis for this analysis. (We will of course review the continuing flow of questionnaires coming to us in case the picture changes, but past evidence has shown little substantial change at this stage.)

Our 1995 sample population
Of respondent authors, 51 per cent had penned more than ten articles each during their academic and professional career (1991 = 61 per cent) including 35 per cent with 20 and more (49 per cent). Twenty-nine per cent had written five or less (20 per cent) including 8 per cent only one (4 per cent). These comparative data show that a less "mature" author sample has responded than in 1991, but nonetheless still reflect well the judgemental profile of academic publishing across the whole Literati Club. Forty per cent were UK based, 31 per cent from Canada and the USA and 7 per cent from Australia and New Zealand. Overall, authors from 55 countries replied.

Among authors with more than ten articles to their credit, only 39 per cent had written more than one for MCB University Press (no 1991 comparative data). This statistic and the share MCB University Press held of all other authors' total written output, would indicate respondents were well aware of publishing processes at large, and not at all constrained by any dependent relationship with MCB University Press *per se.*

As in 1991, authors' primary areas of academic interest ranged across the whole MCB University Press portfolio, with no single area involving more than 8 per cent. Forty-six per cent of authors had an article in preparation or nearing completion that they would like an appropriate MCB University Press editor to consider, which was a significant increase on 1991 (28 per cent). Seventy-seven per cent of authors made use of an academic library to prepare their articles (62 per cent), the balance normally using corporate or public libraries – reflecting a more academic sample basis than previously.

With whom and how do we compare?
Once again, as in 1991, well below half the authors felt sufficiently competent to suggest which publisher MCB University Press was most like. But, once again, their choice was Elsevier at 15 per cent (14 per cent) followed by Pergamon at a similar 10 per cent on both occasions. Wiley, Cambridge University Press and Blackwells were each chosen by 7 per cent. Continuing this comparative vein, 46 per cent described MCB University Press service to its authors as superior to other publishers with whom they had worked, 51 per cent as on a par and 3 per cent as inferior. This question had never been previously asked, but another was in 1992 on acceptability of service during article submission ($n = 98$). The 1995 data showed that 68 per cent were fully satisfied (70 per cent), 24 per cent were satisfied (28 per cent) and 8 per cent were convinced it could have been better. Noting the 1992 sample size, this cannot be seen as significantly better or worse statistically, but the important point to mull over is that no overall perceived improvement has been registered – despite considerable efforts with our editors and Literati Club guidance notes and workshops. There may well be good explanations but it cannot yet be called competitive progress.

Room for improvements
The same messages as in 1991 were repeated by those who were disappointed with the service they received as authors. Although they remain a small minority their concerns are real and have not decreased. Some editors and reviewers on some occasions are slow in turnaround, less than constructively critical in feedback, and journal publication lead times, even when an article has been accepted, can seem interminable (one respondent reported five years!) and intelligence on likely publication dates elusive. The conclusion must be that, to address these residual problems, MCB University Press needs to think laterally – and it is very much a part of our electronic publishing strategy to do just that. Transparency of the whole submission-acknowledgement-review-revision-scheduling-proofing-publication process is clearly within our grasp. Authors specifically suggested the widespread use of e-mail, and electronic peer review.

The overwhelming good news carried the opposite message – Literati Club was a truly unique contribution and help for authors, articles flowed through MCB University Press smoothly and editors and reviewers did a professional and creative job. The complimentary multiple copies to authors were appreciated. One author eulogized: "MCB University Press is most like Harvard University Press. It strikes a good balance between professional and academic approaches. It provides outstanding professional service without losing the human touch. Frankly, I do not see any room for improvement. I received courteous and personalized attention throughout". Another commented: "You're in the vanguard. Always on the trail for quality and professionalism, but willing to publish provoking articles from newcomers in the field".

The prices MCB University Press charges for its journal subscriptions were again criticized by a small proportion of authors. They advanced the case that if they were cheaper more libraries would be able to stock them and, as such, wider readership would occur. Once again, we are determined to deploy the arrival of electronic publishing and networking to think laterally on prices as our cost structures metamorphose. However, the real scope for gain in the electronic age seems most likely to arise from the greater potential for easy and broadened access to, and therefore greater readership of, the knowledge that has been purchased – i.e. very much greater value added for/derived from the subscription paid. The introduction by MCB University Press of CD-ROM archives with 84 of its printed journals (EMERALD), inclusive with their annual subscription from 1996, is an excellent case in point. The life value of all earlier contributions is enormously increased, the holding can be networked extensively, and the space required to hold back issues in libraries can henceforth be used more effectively. It is also further exemplified by the MCB University Press Internet access for subscribers (on URL http://www.mcb.co. uk) to forthcoming contents pages and abstracts, conferencing precursors for journal special or dedicated issues and opportunities to follow up ideas with authors via e-mail services and editorial forums.

The costs of capture and presentation *per se* do not seem to offer much scope for further savings other than printing, and printing at the publisher node must be expected to run in parallel with solely electronic delivery plus the opportunity to print on demand at the reader node, for some considerable time yet. During the autumn years of the Caxtonian framework, however, we are only too well aware that we have been able to make no significant progress other than modest price economies for niched new launches to key customers, and for subscribers willing to forgo their personal copies of publisher node printed journals (Rowley, 1995).

Understanding authors' use of and attitude towards electronics

The rationale which led MCB University Press on to the Internet, and has since positioned us as one of its main academic publishing activists, was quite simple. It clearly offered both major threats and opportunities to us as publishers, and we could not wait and see. We resolved to get in among it, to pioneer new ideas that took maximum advantage of its value-adding contribution to academic publishing. In short, we decided to embrace all the opportunities we could find. From the outset, this included our communications with our authors and editors and, as such, Literati Club joined the Internet in September 1995. The occasion of the membership survey early in 1995 was taken to gather as much understanding as we could of our authors' and editors' existing electronic habits and their attitudes now and to its future.

The first piece of intelligence we possessed had been collated prior to our acceptance, from 1993, of authors' articles on floppy disk. Previously, of course, all articles were re-keyed with the opportunities for errors and high demand for proof checking. In 1992, 57 per cent of authors had agreed it was reasonable to require disks from them – and we moved immediately to that policy. Such is the pace of change in electronic publishing, however, that although we are most grateful for author collaboration with us, scanning of articles rather than disk transfer and reformatting now runs at over 75 per cent of all articles.

The second piece of intelligence covered the critical decision to take Literati Club on to the Internet. During 1994 we sought to find out how many authors were making use of e-mail. The replies we got ($n = 321$) showed that 54 per cent were active and this rose to 64 per cent among human resource specialists. As will be readily perceived, any total move to Internet would imply loss of contact with nearly half our membership – yet nevertheless we did it. The cost benefits of the Internet approach beckoned irresistibly and we knew we were going with the grain. Microsoft 95 was soon to be launched, and the global media hype was moving everyone and anyone to conclude that they ought soon to come to terms with the Internet. We resolved to devote all our resources to seeking to make the Literati Club site as valuable as possible, in the context of the whole of MCB University Press developing the site too.

Our full-scale 1995 study of authors has borne out the e-mail data, and further shown that 47 per cent are making use of Internet conferencing services too. Australia and New Zealand led the way (62 per cent), followed by Canada and the USA (57 per cent), then the UK (41 per cent). Australia and New Zealand again led (49 per cent) in volunteering to join MCB University Press' own Internet conference programmes as convenors and key contributors when they were launched in September 1995, followed by the UK and Canada and the USA, both at 31 per cent. While human resources remained an important area for such activities it was joined strongly by strategic management and information management. Fifty-eight per cent of Australian and New Zealand authors were already members of an organization-wide electronic information network, and 40 per cent of their libraries had a designated individual concerned with networking. They are the world leaders. The comparative figures for the world at large are 38 per cent and 25 per cent; for Canada and the USA, 40 per cent and 21 per cent; and for the UK, 35 per cent and 23 per cent.

We referred earlier to our intuition that taking Literati Club on to the Internet was working with the grain. We were gratified that 83 per cent of authors, at

August 1995, agreed that: "In the not too distant future electronic highways will play a much more significant role in publishing". The figures were 98 per cent from Australia and New Zealand, 85 per cent from Canada and the USA and 81 per cent in the UK.

The form in which electronics is expected to have an impact was researched further using a six-point semantic differential scale ranging from unlikely (= 0) to highly likely (= 5) for a series of scenarios over the next five years. The findings were broadly unaffected by the author's maturity in terms of articles written, or by the major countries where authors reside. Table I shows the overall distribution of feelings for each statement.

It can be seen that the most strongly held view (av = 3.85) was that paper-based journals will complement their services with electronic conferencing and CD-ROM. We are delighted at that since it is our pre-eminent approach. It was closely followed (av = 3.82) by the view that CD-ROM searching will become the norm, although online searching of publishers' records (av = 3.2) would appear at first sight to be contradictory. We envisage that online searching mainly for the most recent materials, and then linked to networked desktop PCs. This latter scenario was also well supported (av = 3.6).

The event seen as least likely by the authors (av = 2.02) was the advent of electronic journals to replace printed journals. The UK pulled this average down (1.8) with Australia and New Zealand ahead again (2.6) and Canada and the USA at 2.1. This did not surprise us although we believe such publications have a great future when linked to publication on demand shielded by encrypted access. They seem to be the evolving destination of the mixed media scenario seen as the most likely in the next five years. With effect from 1996, MCB University Press is offering six of its journals either electronically or paper-based with a subscription saving of 25 per cent if taking the former. By 1997, it is hoped to publish all articles on the Internet immediately they have been appropriately formatted for the journal that has peer reviewed them and thereby accredited them.

	0	1	2	3	4	5	Average valence
			Percentages				
Paper based journals will complement their services with conferencing and CD-ROM	0	3	5	22	44	26	3.85
CD-ROM searching will become the norm	1	3	11	17	34	34	3.82
Majority of current awareness on networked desktop PCs	1	3	10	28	37	21	3.6
CD-ROM replaced by online searching at publisher	2	9	18	30	28	13	3.2
Electronic journals will replace paper-based	17	25	21	19	14	4	2.02

Note: [a] Semantic differential runs from 0 = unlikely to 5 = highly likely

Table I.
Attitude valences towards electronic publishing futures
($n = 814$)

We see this latter approach as offering greater satisfaction to authors and readers alike as it will afford virtually immediate publication rather than waiting in line for space in a forthcoming paper-based issue or going through the time-consuming printing and mail delivery processes. If this can be achieved, and the process of peer review, comment and feedback can be equally made more expeditious, we believe we can make good inroads on the lingering problems which well below 10 per cent of our authors were sadly still reporting five years after we launched the Literati Club.

What we know; what we don't know

This article has analysed our leading learning activities in the Literati Club since its foundation in 1991. It can be seen that our drive to build supportive relationships has been noticed and that we are recognized as superior in our service to authors by a most pleasing margin. Nevertheless, we have failed to make any serious inroads into the problems encountered by under 10 per cent of authors in their editorial submission/publication sequence.

Although electronics are never going to be a panacea for these difficulties, they can give a transparency to the process which will enable the publisher to identify problems before and as they arise, to offer support to editors to overcome them. The editorial processes, conducted outwith MCB University Press, require considerably more focus in the coming years than we have given in the past. Even where we have been in direct control – i.e. publishing/proofing interfaces with authors – we have not advanced the standard of our performance, though it remains extremely high. The continuing search for solutions to these challenges is, we believe, the proper foundation for a high quality and effective evolving relationship with all our authors.

At the same time, these hygiene factors in publishing cannot divert us from the paradigm shifts available in electronic publishing that will enable 100 per cent of authors to gain better and more rapid dissemination of their work. Internet clearly affords opportunities for prepublication discussion and review, follow-up with interested readers and the low cost creation of a myriad minority themed special issues/colloquia/journals all printed on demand by the reader.

Every future we explore, however, seems to need structure if the sheer volume of data is not to overwhelm us all. Since Caxton, it has been the publisher and the printer who conceived the artefact of a journal, appearing in several issues each year, which gave structure. That artefact is currently being deconstructed along with all that goes before and after it. The new artefact will be a quality stream of continuous, accredited knowledge and information, classified imaginatively in the expectation of subsequent search, but brought to our attention, selectively as and when we are known to have a germane interest. Our close colleagues do it all the time face to face, in the corridors and meeting places. The renaissance publisher in the electronic age will do no less, and authors will want to be working with that publisher. As yet, we are only dimly aware of how precisely we can achieve it, but we are working on it in partnership with our authors. There is no other route we can discern that has any chance of seeing us there on the far side.

References

Argyris, C. (1993), "Education for leading-learning", *Organizational Dynamics,* Vol. 21 No. 3, pp. 5-17.

Brown, R. (1993), "Write right first time", *Literati Newsline* accessible via Literati Club on http://www.mcb.co.uk

Davidow, W.H. and Malone, M.S. (1993), *The Virtual Corporation,* Harper Business, New York, NY.

Day, A. and Peters, J. (1994), "Quality indicators in academic publishing", *Library Review,* Vol. 43 No. 7.

Day, A. and Toledano, K. (1995), "Quality inside out and outside in", *The TQM Magazine,* Vol. 7 No. 2, pp. 23-8.

Evans, P. (1994), "The peer review process", *Literati Newsline accessible* via *Literati Club* on http://www.mcb.co.uk

Jordan, T., Bruce, B. and Wills, G. (1994), "Realizing the benefits of a marketing intelligentsia", *Marketing Intelligence & Planning,* Vol. 12 No. 6, pp. 21-34.

Rowley, J. (1995), "Issues in pricing strategies for electronic information", *Pricing Strategy & Practice,* Vol. 3 No. 2, pp. 4-13.

Sandelands, E. (1995), "Which journal? The politics of where to publish", accessible via Literati Club and ANBAR Electronic Intelligence on http://www.mcb.co.uk

Wills, G. (1992), "Enabling managerial growth and ownership succession", *Management Decision,* Vol. 30 No. 1, pp. 10-26.

Wills, G. (1995), "Embracing electronic publishing", *The Learning Organization,* Vol. 2 No. 4, pp. 14-26 (see Chapter 2).

Chapter 6

Rogue learning on the company reservation

Introduction
Individual and corporate learning
How can individual learning be turned into "corporate learning"? This is a matter which concerns both participants in staff development programmes who want to see their innovative ideas adopted, and their sponsors who want to see business benefits result from their investment. The extent to which staff learning is corporately adopted depends on a variety of factors, not least, of course, the quality of the new ideas which emerge in the wake of a staff development programme. Three further factors are key.

The first is the political and personal power skills that managers need in order to have their proposals adopted within an organization. This aspect of a manager's development has been explored in the present author's book on self-development[1]. Middle managers need to be able to "lead from below the top" and "succeed in an imperfect organization"[1, pp. 230-2].

The second factor is that staff development programmes should meet business and corporate objectives. This is the approach of many project-based development programmes, for example, those based on "action research" or "action learning". However, as Critten has persuasively shown[2], unless supported by an environment in which learning and development are encouraged, this approach may be insufficient. This point leads on to the next factor.

The third factor is the organization's attitude to proposals for change coming from staff – how far they are encouraged to propose corporate level initiatives and whether there are mechanisms by which staff can have their proposals considered. This notion of designing organizations which are generally receptive to ideas for change emanating from staff is a theme that has long pervaded writings on organization and industrial participation, dating back at least to the classic contrast of "mechanistic" and "organic" organizations[3]. Contemporary manifestations are found in concepts of "learning" and "transformational" organizations (e.g.[4,5]).

This article reports research which brings together all three of the above factors. It looks at the effects of a series of staff development programmes that

The research reported in this article was conducted in the course of Tom Reeves' tenure as International Management Centres' 1995 Revans Professor and thanks go to IMC for its sponsorship which made the research possible. Thanks also go to senior management at MCB University Press and at Seagram for agreeing to allow their companies to be exposed to scrutiny, and to the individual members of staff within those companies who spoke freely about their experience of action learning.

By Tom Reeves,
The Learning Organization,
Vol. 3 No. 2, 1996

formed part of a sustained attempt to cultivate an organizational climate conducive to innovation, in which staff produced proposals for corporate change while being developed in the skills needed to deal with their organization.

Honest rogues
Individual members of an organization are rarely, if ever, allowed unbounded licence to implement change. They operate perforce, in the metaphor of the title of this article, on a "company reservation", the bounds of which are tacitly set by all the managers involved in approving new initiatives. Any would-be initiators who might be tempted to stray over the permitted bounds – potential "rogues" – are held back by their knowledge of what is and is not permissible within their organization.

The term "rogue" has many connotations, some favourable, some not so favourable. All of which depend on the perceiver. Here, "rogue" is taken to mean someone who pursues a personal agenda for action based on his or her alternative view of the organization's best interests, the directions it should take and how it should be run, and who does this with disregard for the formally agreed or authorized corporate agenda. Such persons, although perhaps appearing subversive to the organization's power-holders, may well not see themselves as subversive, since they believe that they have the organization's best interests at heart. Rogues may appear particularly dangerous if they stray beyond the company's bounds of acceptable behaviour. But, within the boundaries of the metaphorical reservation, rogues, although disruptive, may be seen as a creative, constructive force. In fact, an element of roguery, of intellectual non-conformity, is probably essential if radical initiatives for change are to be forthcoming.

The research
Corporate impact of action learning
The research to be described in this article resulted from the author's appointment to be International Management Centres' Revans Professor for 1995. The task of the Revans Professor is to carry out a piece of research that will advance the practice of action learning. The aims of this research reflect the interests of both the author and International Management Centres (IMC) which, through their action learning programmes, precipitate considerable numbers of proposals by staff for organizational change.

It was agreed that the research should make a comparative evaluation of the corporate impact of action learning programmes in two companies; MCB University Press, publisher of this journal and sister company to IMC, and Seagram, the international drinks company. Both companies have made extensive use of action learning programmes, though in very different ways. MCB University Press (MCB UP) has systematically used action learning as an instrument for developing its business over many years; Seagram has made more sporadic use of action learning, primarily as a means of individual staff development.

The research was to explore how far it had proved possible in these two companies to enhance organizational learning through the use of individual action learning. Its key aim was to develop understanding of the process of deriving corporate development from individual learning, and to consider how this might be extended.

Action learning

Action learning requires people to tackle a challenging problem or task in their work, and learn from the experience. It is a somewhat unpredictable process. It cannot be known in advance quite what people will gain from their experience, nor exactly how, if at all, the organization will benefit from the project. Action learning, by involving its participants in corporate level projects, typically raises their expectations and aspirations. Weinstein has described action learning as subversive: "it values everyone...it stresses questioning...it insists on actions...it examines everything"[6].

While undertaking their learning project, action learners are normally members of a small group of other action learners who support and encourage one another through the vicissitudes of their inquiries, action and learning. The discussions and reflection in these groups or "sets", can cause participants to question current ways of doing things. They can develop fundamentally new, perhaps highly critical, understandings of their organization and their place within it. As they grow in confidence, as tends to happen on action learning programmes, they seek outlets for their new-found capabilities and ideas.

Both from the point of view of an employer wanting to develop staff to perform a particular job, and from that of an individual seeking to make him or herself more effective, action learning clearly has rogue propensities. This has not, however, deterred many employers who regularly send their managers and other staff on action learning programmes, nor staff from attending them.

Indeed, within limits, the rogue elements of action learning are usually seen as something to be welcomed. It is valuable to have middle and junior managers seeking greater responsibility or autonomy, who are not afraid to challenge the status quo, and who are willing to initiate change from their level in the organization – all ingredients of what has come to be known as the "learning organization".

Corporate development

Action learning combines profound and lasting individual learning with the organizational benefit of a useful project. But can it do more than this? Can action learning be made to yield corporate benefits which go beyond developing the capabilities of individuals with projects undertaken as a spin-off? Can action learning programmes be used to develop an organization as a whole, using the individual learning as a platform for developing a way of operating or a culture which promotes wider corporate learning? These are the kinds of questions that led to the present research.

Again in the metaphor of the title, can action learning's rogue propensities be tamed and harnessed to corporate goals? Can the corporate good be served by allowing action learners to roam the company reservation freely?

Before addressing these questions, however, it is first necessary to give some further background to the research and the companies in which it was conducted.

MCB UP and IMC

MCB University Press is in the business of publishing academic journals and has grown from small beginnings to being possibly the largest management journal publisher in the world. It is nevertheless quite a small company employing about 190 staff.

MCB University Press has invested considerable sums over many years in action learning programmes offered by its sister company, International Management Centres. Managerial, professional and administrative staff have all participated in these programmes, obtaining qualifications such as an MBA, Diploma or Bachelor's degree. MCB UP has also regularly mounted non-qualificatory management development programmes based on action learning. Both types of programme have had the intention of developing the company's business as well as its staff[7].

Professor Gordon Wills, principal of IMC and chief executive of MCB University Press, in a book about these and other IMC programmes, has suggested that using action learning to develop in-house experts who can then develop others can turn a company into a form of school[8]. He has also reviewed the possibility that MCB University Press might have developed into a learning organization[9]. MCB UP has repeatedly adopted the recommendations of individuals and groups which have undertaken action learning projects, with numerous resulting changes in MCB UP's policies and operating procedures.

But whether or not MCB UP measures up to an ideal type of learning organization is primarily of academic interest. The more practical issue is whether MCB UP could be more effectively harnessing individual or group action learning to corporate ends. What kinds of corporate contribution are feasible and how can they best be achieved?

These are questions of concern to any company investing heavily in staff education and development. Developing people's capability may have only marginal effects on corporate performance. Are there more effective ways in which individual development can be made to contribute to corporate development?

Seagram

The company chosen for the purpose of comparison was the international drinks company, Seagram, which, over the years, has put many of its staff through the IMC MBA programme, and has recently put a cohort of staff through IMC's BA administration programme. In the event, Seagram proved of less value as a basis for comparison with MCB UP than had been anticipated. Nevertheless, it afforded some useful parallels and contrasts.

Method of research
Individual interviews were carried out during the summer of 1995 with senior managers in each company, and group discussions held with managers and others who had undertaken, or were currently undertaking, an action learning programme.

These interviews and discussions were relatively unstructured, in order to allow relevant topics that emerged to be discussed. They did, however, focus broadly on the following:

- the outcome of the action learning for the individual and for the company;

- views on the organization, its culture and effectiveness;

- opportunities for taking initiative, contributing ideas, feeding back experience, sharing learning and getting one's recommendations heeded.

Within MCB UP, four of the directors were interviewed and five group discussions held with representatives from a current MBA set, a SEMAP (senior management programme) set, a bachelor's degree set, and those who had earlier completed an MBA or bachelor's degree. In all just over 30 people were interviewed or participated in a group discussion.

Within Seagram, the president of the division which had used IMC's action learning programmes and its human resources manager were interviewed, and a group discussion held with five administrative staff who had recently completed a bachelor's degree. It had been hoped also to interview some managers who had earlier completed an MBA. However, many of these have since left the company, while others are posted abroad. The few it might have been feasible to interview declined.

Key findings
Bounds to autonomy
A recurrent theme of the research was the apparent tension within MCB UP between action learning's encouragement to expand the scope of people's thinking and activities and what was believed to be impermissible behaviour within the company. Informants – at all levels – talked about, on the one hand, the autonomy provided by the action learning culture and, on the other, perceived bounds to inquiry and action. There were, of course, other issues raised and these will be discussed, but the reconciliation of these conflicting messages was dominant. It is this tension which the title of this paper attempts to encapsulate in its metaphors of rogue learning and a company reservation – a space within which people had a licence to apply action learning principles.

It needs at once to be stressed that virtually everyone interviewed seemed to be quite adept at handling the tension. Despite new aspirations and new capabilities, they had accommodated their evolving selves to the constraints, which were largely accepted as an inescapable aspect of working in an organization. They were far from being rogue learners. They well understood that to survive and be successful, their new-found capabilities and aspirations had to be contained within the company reservation. Rogue learning was thus domesticated.

Action learning culture
A second key finding was that within MCB UP a climate or culture had evolved in which many of the key aspects of action learning – its emphasis on questioning and inquiry, on projects and change, on openness of discussion and debating decisions on the basis of evidence – could be safely manifested, indeed were encouraged. Within its bounds, and even in the most liberal of organizations there inevitably are bounds, the company reservation was a place where one could overtly practice an action learning style. One could be open, ask potentially awkward questions, expect collaboration, criticize present ways of doing things, put forward ideas for change, be allowed to make mistakes and, indeed, exercise a considerable degree of autonomy. Action learning principles and practices pervaded MCB UP and have shaped the corporate way of doing things. The selected quotations shown give a flavour of the nature of this culture.

MCB UP's action learning culture
Action learning culture is represented all the way through the company – in meetings, relationships, etc. It's actively encouraged. In one's day-to-day job, action learning happens. You're left to discover the answers, look at alternative ways of tackling problems. It slows the process but you get a richer outcome.

There's a language, jargon, acronyms. It (the learning culture) nibbles away. Not leaps and bounds. The company's financial stability, the certainty of profit, gives confidence to experiment. We're cash rich – at least richer than most. Action learning prevents complacency. It's helping us to stay ahead of the field. There's also the academic influence – papers are thorough, nothing is overlooked.

Board members involve people at bottom level in the thinking process behind projects. It's not unusual to get faxes from a director. But several people can be involved without co-ordination. However it seems to work. There are different agendas which are not always pulled together. But if something was felt to be going off the rails it would be pulled together by academic debate. Eventually we would reach consensus.

If you have a problem you go and seek a solution – every day.

You get such learning in the company – you have to find out. You move around. You survive by finding out. There are few hostile barriers. People are not threatened by you asking.

The owners will back you if they feel you are right. We pull together in a logical way. And there's always an eye on the bottom line.

Projects wouldn't be done without SEMAP (compulsory action learning programme for managers). It takes you into other areas. You wouldn't go into other departments otherwise and ask them why or how they do things...You can't do anything on your own in this company.

The action learning culture is 'Ask the right questions and act on the results'. The majority of people feel they can ask questions and have a go.

In MCB you're not given the answers – you're encouraged to address the issues.

The action learning culture is so strong, you sink or swim. There's little allowance for people just getting on with their job and not going on any of the programmes. There's no stigma if you don't, but you're ignored. Squeezed out. At a certain level it's expected that you'll do an action learning programme.

Motives for introducing action learning

The degree of corporate impact of action learning in the two companies was largely determined by each company's motives for using it, and the extent to which top management encouraged its manifestations. There was little evidence of action learners acting as a force for corporate change other than within parameters set by top management.

In Seagram, it had never been intended that action learning should be a vehicle for corporate, as opposed to individual, development. The divisional president, a PhD holder, has had a lifelong interest in continuing education. IMC's route to educational qualifications for employed staff through action learning filled, as he saw it, an individual development need. In an article describing an earlier programme he had sponsored in the Seagram-owned International Distillers and Vintners Company, the action learners were said to have had a catalytic effect on others in the company[10]. However then, as now, the benefits to the company were seen primarily in traditional terms: having more competent, knowledgeable and effective staff, and having actionable recommendations. Thus, while the action learning projects could be expected to result in operational improvements, the rationale for using action learning was essentially one of promoting individual rather than corporate development. This intention was reflected in the benefits cited by Seagram informants shown below. The focus of discussion was exclusively on the benefits to themselves as individuals, and the difficulties encountered in capitalizing adequately on these benefits within the Seagram structure.

A further factor inhibiting a wider corporate impact in the case of Seagram may have been the dispersal of the action learners throughout the company; there appeared never to have been a significant number of them in any one place at any one time.

Cited benefits of action learning in Seagram

The BA (Bachelor's Administration degree) has given me a much broader outlook. It has changed me as a person.

The projects built our skills. They were relevant at the time, but we've since moved on – you need to know when to leave something. Although the project wasn't personally relevant, it enabled me to use skills I learned doing it – assemble facts, market research. I'm still putting them into practice…

I spent hours! I did something one wouldn't otherwise have done. I wouldn't have had the time without being made to do it. I chose my project out of interest, and it became even more relevant.

A traditional degree feeds you with knowledge. This degree was not so much about the quantity of knowledge, but attitudinal – how you worked, and how to make things work in your favour.

The benefit for me was learning how to learn. How to glean information.

You can go away and apply action learning. It's common sense, an attitude. This degree gave me a good attitude. I'm not frightened to ask questions. I'm able to show ambition, commitment, prove myself, able to fight – a big spin-off.

Force for change
MCB UP, on the other hand, did intend that its use of action learning should have a corporate as well as an individual impact. This has undoubtedly been achieved: action learning has been used in a way which has been profound and far-reaching. Virtually all managers and administrative staff have been through an action learning programme of some kind, if not one leading to an IMC qualification, e.g. an MBA, then almost certainly one of MCB UP's regular management development programmes which are compulsory for certain managerial grades.

This extensive exposure to action learning processes and principles creates the possibility that the action learners themselves might have become an independent force for corporate change. There was, however, little evidence of this having happened.

Before conducting this research, the author had thought it possible, if a sufficient number of people in an organization were exposed to the questioning and challenging ethos of action learning, for these action learners to act as an independent force for corporate development and change. However, in the test case of MCB UP, where all managers and many administrative staff have been through at least one action learning programme, action learning did not appear to be having this kind of autonomous impact.

Action learning had, nevertheless, had a profound impact on MCB UP's culture and style of operating. However, as far as could be ascertained, this was only in ways which reflected top management's intentions and objectives. There was no licence for action learners in MCB UP, nor in Seagram, to take unauthorized initiatives, and these tacit bounds appeared to be understood and accepted.

The corporate impact of action learning in MCB UP would appear to have resulted almost entirely from top management's intentions for it. The motives for making use of action learning in MCB UP were complex. In large part, its use originated in the fact that the company's founders were academic management specialists who were already convinced of the advantages of action learning as a means of developing managers. Founding a company offered an opportunity to put theories in which they believed into practice, and indeed it was said they felt a moral commitment to do so.

However, the rationale for the use of action learning in MCB UP is far broader than this. As stated by one of the directors, the aims of action learning for MCB UP were, and are, to:

- build a business – with more capable people;
- grow from within, promoting people internally wherever possible;
- increase turnover;
- control costs;
- avoid the use of external consultants – through the action learning projects MCB UP's own people make the recommendations for change;
- promote collaborative networking within the company.

There were several indications that these objectives have been realized. The company has grown and been financially successful. The number of projects which have had their recommendations adopted demonstrate the company's commitment to innovation in marketing and operations, and generally to sustaining continuous improvement. The action learning projects provide the occasion for these changes, while the repeated action learning programmes sustain the impetus. It is difficult to see how so much innovation could be achieved by other means.

MCB UP's owners have had sufficient confidence in staff capabilities to have pursued a policy of promotion from within. Several staff have risen through the ranks to director level; all of them had been through at least one action learning programme, most through several, and they were willing to attribute their advancement at least in part to the way their action learning experiences had enhanced their capabilities. Other staff who had been through the action learning process similarly testified to their enhanced capability. Recurrent themes were growth in confidence, greater knowledge of the company and its business, an ability to tackle problems, the acquisition of a new outlook and new ways of going about work tasks. Unlike Seagram, there was a strong emphasis in the discussions on how new capabilities could be put to use for the benefit of MCB UP.

Indeed, there was a marked absence in the MCB UP discussions of "barriers" or "blockages" to authorized action. Once a recommendation had been approved, people seemed to be able to find wide support for its implementation. Talk was more of autonomy than of constraints. If there were barriers, they were seen as something for the project holder to overcome rather than externalize and blame on others.

Boundaries
A corporate culture has built up in MCB UP and has several times been referred to as the company's "action learning culture". The nature of this culture has been indicated earlier in this article; it included problem solving based on inquiry, collaboration across company networks, openness in sharing information, and debating decisions around evidence. The action learning culture is both an ethos and a way of doing things. It straddles all levels, and is actively supported by top management, who appear conscientiously to strive, in accordance with the spirit of action learning, to be receptive to ideas for change. Using the metaphors of the

title again, they have created a relatively safe reservation on which action learners can roam and apply their action learning principles.

However, as noted earlier, the reservation has its boundaries. Some of these relate to the owners' sense of prerogative regarding key business decisions. Several informants, in this context, said there was a power culture, reflecting the domination of the owners. Some boundaries, according to certain informants sensitive to the human factor, were believed to derive from owners' and top managers' natural resistance to criticism and challenge – problems with which several felt good action learners should know how to cope. The boundaries of the reservation were not always clear, indeed were sometimes only surmised, as the comments below show.

The correctness of the informants' perceptions expressed could no doubt be questioned. The point, however, is that within the context of an action learning culture some people felt it was inappropriate, indeed risky, to speak their mind too openly or attempt to do certain things.

The sense of barriers and boundaries in MCB UP

There *are* barriers. They need to be removed...But you can speak out here. I wouldn't want to work elsewhere. The scale of the problem compared with other firms is miniscule.

I'm more comfortable myself knowing the bounds. We're told there are no bounds. But they are there – invisible. Sometimes you make a decision, but you're uncertain whether you're transgressing a senior manager's view.

It's not OK to step on toes, to uncover things that are not OK. It's difficult to challenge established practice...

This group then debated the political consequences of making "untoward" criticisms, concluding:

It can damage your career if you step on too many toes.

The resistances to change (of senior managers) is not dissipated by going through action learning processes. They're not sitting there with a belief that because it's action learning it's OK. It affects them personally. It questions their judgement.

They tell you to be open, but you've got to recognize that you're dealing with people.

I certainly feel I have questioned too much...I want to do different things and get frustrated because I can't. I want to question things. If you don't get satisfaction then there's frustration. It can build up to such an extent you ask is it worth doing?

[In carrying out our projects] we did stir up some hornets' nests. You had to be careful what you asked. We got some cagey replies. But it was the only way to do it – not to mind if you trod on toes...If you're doing a project it implies that someone is not doing what they ought to be doing.

Certain projects were said to be out of bounds:

The major processes of production and publishing tend to stay the same, with the structure changing around them. You can experiment in marketing as long as you don't touch the core. We don't tamper with the core of the business. The business formula we've got works.

You can try to take a project too far. It's easy to get involved in the politics and then out the door! No one was actually directly sacked from the course. But it added to someone's downfall.

Discussion

Action learning's impact

One purpose of this research was to explore whether the use of action learning could lead to some kind of corporate development which was more than the personal development of the individual action learners.

MCB UP, unlike Seagram, had deliberately sought to use action learning as a basis for establishing a particular corporate culture and style of operating. The kind of corporate development that might be expected to result from doing this would be for business and operational developments to be shaped continuingly by new projects, and for the company's management style to take on characteristics which reflected the ethos of action learning. A positive outcome would be for management style to become more open and collaborative, for new ideas and proposals for change to gain a more ready hearing from senior management, for them to be widely shared and discussed constructively, and for there to be a high degree of "decision making by debate" rather than by command. A less sanguine scenario, of course, could be a growth of dysfunctional activity, for example, defensiveness and secrecy in the face of questioning and inquiry, or the pursuit of divergent priorities and directions with consequent dissipation of effort.

MCB UP seems to have achieved the positive benefits. It has established a company-wide collaborative network, it has benefited from a considerable number of projects and, critically, it has created a corporate atmosphere in which experimentation and learning can flourish.

It would be fanciful to suggest that a "learning organization" had been established. Organizations only learn in a metaphorical sense. The invariable reality is that it is individuals within an organization who learn. They may, in the course of their everyday work, have a catalytic effect or pass on their learning to others; or, when project recommendations are adopted, have their learning permanently incorporated in new ventures, practices or procedures. The point is that for individual learning to have a corporate benefit, specific steps have to be taken to make this happen.

Steps to corporate benefit

Some of these steps are well known; for example, ensuring that the learning is relevant, that senior management support its application and providing suitable outlets for people's new capabilities.

On the basis of this research, a number of other steps to enable individual learning to lead to a corporate benefit are as follows.

Create a company-wide climate in which learning and its application is encouraged, expected and supported. In MCB UP this was manifested in the so-called action learning culture. This culture had resulted in part from repeatedly running action learning programmes over many years, and also, initially, from the owners, and then others, manifestly practising the principles of action learning themselves. The success of this culture was evidenced by the complete absence of people saying they had to fight to get a hearing for their ideas or recommendations.

Make induction into this culture a standard part of everyone's learning. In Seagram, it was clearly understood that the action learning was essentially a personal benefit, preparing employees for relevant opportunities should they arise (which they did for some). But otherwise, they were warned, they would have to find ways of applying their learning within their original jobs.

In MCB UP, by contrast, in addition to the personal benefit, action learning participants were also inducted into a corporate culture. They were expected to operate in future on a wider front. This might mean doing one's original job differently, although several stressed that this was not possible, but it did mean one could participate in the general problem-tackling life of the company.

Create specific opportunities for recommendations arising from learning to be heard. A learning culture is not enough to ensure that recommendations arising from learning projects are heard, let alone implemented. Occasions on which learners can present their findings and recommendations for change need to be routine, ideally with senior management present.

Define the reservation
A theme that has run through this article has been the idea that individual learners can most effectively turn their learning into a corporate benefit if they are operating on a "reservation". While the above precepts are all relevant to the creation of such a reservation, there are a number of other ingredients that this research has shown go to make up an effective one. They are:

- a clear licence for taking initiative and action (in the case of MCB UP, this is achieved by the repeated action learning programmes);

- acceptance of investigation, i.e. a lack of defensiveness towards inquiry and questioning;

- scope for taking on more responsibility without promotion or transfer (something which was difficult for the group interviewed in Seagram);

- acceptance, indeed expectation, of innovation from all;

- time for problem solving and initiating action – not only allowing time for this, but also not unduly diverting people with directives to attend to other matters;

- cultivation of an ethos that people can roam free (within bounds), that they are not "tethered" to their present job description;

- setting some boundaries to the reservation – this was done implicitly in MCB UP, possibly the most appropriate way despite the angst it caused in some. These boundaries need to go wider than the limits of a person's everyday duties, but not be so wide as to allow autonomous action where there should be a collective responsibility or where top management wishes to retain a prerogative.

Bridge from individual learning to corporate benefit
MCB UP has created a culture of co-operation, inquiry and innovation, which contributes to business performance and provides benefits in the form of projects driven by this culture. They thus appear to have successfully harnessed individual learning to corporate ends.

Could similar outcomes be achieved without the use of action learning and its associated projects? Action learning was so instrumental in generating corporate benefits for MCB UP that it is not easy to imagine an alternative scenario to the use of learning sets and, more particularly, projects. Indeed, for MCB UP's particular objectives there may well be no alternative to action learning.

Broadly, there seem to be three possible alternative approaches. One is simply to leave individual training and development to have such effects on corporate development as may spontaneously result from subsequent individual initiatives. This is the traditional approach, widely reputed to be of limited effectiveness.

A second approach is for top management, by its pronouncements and actions, to demonstrate that staff have a clear "licence" to initiate and experiment, and overtly to support the exercise of this licence. This approach complemented the use of action learning in MCB UP.

Amplifying learning
A third approach would be actively to support individual learning with a further programme designed to amplify its effects. Such an "amplification programme" would need two thrusts.

The first would be to develop in individual learners the organizational management skills they need to in order to apply their learning to practical corporate purposes, for example, along the lines delineated in Reeves[1]. In MCB UP this was done through the action learning programmes, but other means, such as workshops specially designed to develop organizational effectiveness skills, could be used.

A second thrust would be to review, in a similar way to the present research, the organization's responsiveness to new ideas and initiatives. This diagnosis could then be followed up with appropriate steps to enhance the corporation's ability to harness individual learning, including perhaps running developmental workshops for managers who are in a position to determine the outcome of learners' initiatives.

Organizational support: reservations and rogues
The effective use of individual learning will almost certainly demand the existence of some kind of company reservation, whether explicitly or implicitly defined. One cannot rely on simply harnessing the obviously useful results of individuals' learning, in MCB UP's case the project recommendations. Learners

must also be provided with opportunities to use their learning to contribute to corporate purposes in their own way. This, of course, is restating the well-worn admonition that if staff who have been exposed to development, are to apply their new ideas in their work, they will need a conducive organizational environment. This research has shown how a conducive environment can be constructed.

It so happens that MCB UP and Seagram have made use of a type of learning which has a reputation for encouraging subversiveness. This is also a feature of many other forms of learning; indeed without it important sources of creativity and innovation would dry up. The question is how far actively to encourage managers to push against the accepted boundaries, i.e. in the metaphor of the title, to be "rogues".

Discovering where the boundaries are, when to challenge them and when not to, is an essential part of any middle manager's development, and to have middle managers who can do this is desirable for any organization's well-being. Exposure to rogue learning, whether action learning or other forms of learning that encourage independent thought, will almost certainly develop this necessary capability. The paradox is that a willingness to expose staff to rogue learning and to risk any consequent roguery is an essential part of organizational support for any learning from which corporate benefits are sought.

References
1. Reeves, T., *Managing Effectively: Developing Yourself through Experience,* Butterworth Heinemann/Institute of Management, Oxford and Stoneham, MA, 1994.
2. Critten, P., *Investing in People: Towards Corporate Capability*, Butterworth Heinemann, Oxford and Stoneham, MA, 1993.
3. Burns, T. and Stalker, G., *The Management of Innovation,* Tavistock Press, London, 1963.
4. Senge, P.M., *The Fifth Discipline: The Art and Practice of the Learning Organization,* Century, London, 1990.
5. Banner D.K. and Gagné, T.E., *Designing Effective Organizations: Traditional and Transformational Views*, Sage, Newbury Park, CA, London, 1995.
6. Weinstein, K., *Action Learning: A Journey in Discovery and Development,* HarperCollins, Philadelphia, PA and London, 1995.
7. Gore, L., Toledano, K. and Wills, G., "Leading courageous managers on", *Empowerment in Organizations,* Vol. 2 No. 3, 1994, pp. 7-24 (see Chapter 7).
8. Wills, G., *Your Enterprise School of Management: A Proposition and Action Lines,* MCB University Press, Bradford, 1993.
9. Wills, G., "Enabling managerial growth and succession", *Management Decision*, Vol. 30 No. 1, 1992, pp. 10-26.
10. Espey, J. and Batchelor, P., "Management by degrees: a case study in management development", *Journal of Management Development,* Vol. 6 No. 5, 1987, pp. 61-8.

Chapter 7

Leading courageous managers on

First the bad news. Our publishing enterprise is threatened over the next decade with massive changes among its suppliers, in the formulation of its products and in its channels of distribution. We are owned and led by a team of partners who have built the enterprise to what it is today but have no burning zeal to see it into another quarter century of growth and development – who would at their age? They talk frequently of selling out to realize their capital, an outcome which would surely mean the end of the culture they and we have created and nurtured over the past 27 years. Meanwhile, they clearly enjoy the fruits of their labours.

Now the good news. Our owners know the bad news to be true. They consider it their personal responsibility to enable those who might wish to prosper with and through the enterprise over the next quarter century to address and surely overcome the challenges. In cold hard terms, however, we must come up with a scenario for the owners that is a better offer than selling out their stakes to folk who might/probably would unscramble our enterprise and its culture.

Until ten years ago the owners managed as well as led the enterprise. The rest of us enjoyed the culture rather well because the owners were infrequent attendees at the offices. The enterprise was their hobby, a sideline to their professional work as university management academics. We were not overmanaged, but we did have to produce what was perceived as the critical management information for the owners on a routine basis.

Ten years ago the enterprise started to get noticeably bigger. We moved to new offices and not everyone knew everyone anymore. Today we have £15 million annual sales, and 150 or so full-time staff headquartered in Bradford, England but with three more offices and a host of agents across the world. We buy-in a great many services ranging from authorship and editing to copywriting and promotional lists, printing and shipping. Not big; but big enough not to be small any longer. It was obvious the owners either had to manage it a bit more, or somehow or other get others to manage it a bit more for them. Fortunately, or wisely for us, they decided to get others to manage the growing organization while they concentrated on learning to lead it. Finally, and most importantly for this analysis, they decided to give anyone and everyone within the enterprise the opportunity to develop into managerial roles if they so wished, and to the full extent of their abilities. This was a considerable but not unduly worrisome act of faith and trust for them. Trust was easily given because everyone did still know everyone in the supervisory and middle management areas, and thought well of one another. Faith was a two-way

With Kathryn Toledano and Lesley Gore,
Empowerment in Organizations,
Vol. 2 No. 3, 1994

commitment between a group of six university academics and some 30 administrative and marketing promotions staffs. If the right opportunities were identified and appropriate support was given, they could become whatever it was that would be needed, plus or minus a few specialist skills as yet unknown.

With a few honorable exceptions, management recruitment outside has not been necessary. Together we have built our enterprise to its present state as the largest academic management journal publishing house in the world, with some 130 journal titles. We always assumed we could do it of course. And the reason why it was not especially worrisome to the owners was that they were all management teachers, and believed their own rhetoric. They believed it was readily possible to develop intelligent staff to tackle the challenges of the managerial job at MCB University Press provided they developed their leadership skills as owners. Immediate and sustained results were achieved[1].

The leadership strategies employed
This article will explore in some detail the leadership strategies which the owners have pursued to enable courageous managers to emerge and flourish. It will sound a lot tidier and better thought out than it has been, but the learning processes accomplished require clarity in presentation. We have never used the word "empowerment", but from reading the literature today, including our own new journal called *Empowerment in Organizations*, it would appear we could readily have done so.

The four key strategies have been focused on:

(1) management action learning;

(2) systems development;

(3) mentoring and coaching;

(4) structural change.

At no time have we been an organization that believed it must introduce "empowerment" to overcome any serious malaise. We are unable therefore to identify with much of the literature about mistrust, frustrated line management or fear of giving up power by supervisors or managers[2-4]. Neither can we accept the perspective Hand[5] offers, that management has the authority to improve the processes but is blind to the problems; and the staff have the knowledge of the problems but are powerless to resolve them. Nor do we find his simplistic solution that the use of "teamwork and empowerment can change all this" very much operational use as we make the journey. Each and every enterprise we presume has to find its own resolutions in its own socio-technical and market context, and contextualization is the hard bit[6].

Finally, before we enter into detailed consideration of our four strategic thrusts, we wish to emphasize that we see nothing as absolutely right or wrong with the sequencing of the actions we took. Again it was context that determined how we proceeded. Their apparent separation was normally very much clearer at their commencement than as time has passed. There was no grand plan although we believe that by staying sensitive to key issues for our enterprise we seem to have clearly addressed the crucial themes[7].

Management action learning
Our first major programmes of management action learning took place in the early 1980s. All middle managers reporting to the owners, who were effectively the senior management team at that time, were required to join a programme known as SEMAP I – Senior Management Action Programme. The owners' experience as first-wave UK university business school academics in the 1960s and 1970s had convinced them that action learning approaches after Revans[8] were likely to be the most appropriate for our staff. On the international scene they had, in 1982, already been responsible for the launch of the International Management Centres – wholly devoted to action learning approaches[9, 10].

Each manager was required to choose a critically important issue for the future growth and development of his/her role in MCB University Press, and then undertake a programme of action learning study over nine months – sharing the journey of analysis, evaluation and the conclusions with fellow middle managers in a Set from right across the enterprise. An accomplished outsider from International Management Centres was invited to act as facilitator to the whole process, and eventually wrote it up[11]. No other use of outside resources was made by the owners although the managers concerned made extensive sorties outwith the organization to gain their better understandings. Sutton[11] reports *inter alia* that the owners reserved the right to veto some of the action learning project notions arising – not unkindly, but simply because they were felt to be inappropriate. While Sutton believed this was bad action learning, the owners were reportedly not convinced they need encourage such freedom!

McClelland[12] assures us that by tackling management development in this way the owners were "gaining competitive advantage through strategic management development" in a way few businesses employing less than 7,500 staff in the USA do even to this day. Their most common steps to develop managers are to send them outside the organization, and for the smaller organization this pattern is almost universal. We did not, however, see our approach as "shifting the primary focus", as McClelland suggests, "from individual to organization effectiveness by placing emphasis on the need for managers who can deal with strategic as well as tactical issues". Nor were we undertaking anything as ambitious as "a complete assessment of management skills, knowledge, experience and formal levels of education". Heaven forbid. Rather we saw ourselves as broadening each of our middle manager's understanding of what others did, and doing so, on the basis of action learning projects, that would be implemented to change and develop the way the enterprise addressed the issues for which the managers concerned felt responsible. And the process made us turn to one another rather than the owners for assistance, guidance and support as we went.

The projects were not trivial. For instance, one envisaged merging two disparate production systems, and stated how to do so. The owners said: "Fine. Get on with it". Another looked at a new campaign to improve renewals driven by the new computer systems. Again, the owners said: "Yes, but how about this as well", and set up a continuing taskforce, reconvened annually to evaluate and energize renewals. It is still at work over a decade later with sophisticated lifecycle modeling[13].

The report-back sessions to the owners were held in the appropriate country house in the Lake District. The greatest fears reported then, and still to this day, were the deep individual anxieties of novices making formal presentations to supposedly fearsome management intellectuals! Focused skill development was provided and still always is in such circumstances. But it is always worth remembering that the self-confidence acquired from communications skills development is indispensable to success for every manager in every role.

Tom Watson Senior, founder of IBM, is widely quoted as saying: "Companies must respect individual employees and help them develop their own self-respect. They must never be psychologically abused". How apt that dictum in the context of personal communications skills.

SEMAP I was swiftly followed by our first in-house action learning Advanced Diploma/Bachelor of Administration programme. Some 20 per cent of the full-time staff attended either half or all of its workshops successfully. The full Bachelor programme lasted two years and provided focused education and training for mature individuals across the major functions and in the skills areas highlighted as vital from SEMAP I.

To summarize, the first stage of management action learning was straightforward, with hindsight. The owners required us to action learn our way forward into the key issues affecting our role in the future in debate and discussion with colleagues right across the enterprise. Follow-through was with focused knowledge and skills development at Bachelor level. All this was followed by the active encouragement of all the emerging senior managers to follow action learning MBA programmes, and to date, around the world more than half of them have done so. Two are now proceeding to their more advanced action learning DBAs (see [14]).

So much general education and training frequently gave rise to requests for professional skills development within the several areas of the enterprise. This was always encouraged not instead but as well. Any reasonable request was accepted. A company-wide supervisors' programme was run using external facilitators. A cadre of marketing services staff spent a year on the British Direct Marketing Association's Diploma programme run at the University of Bradford. Computer database marketing training, training for new production systems and in finance, and full-scale computer studies over four years in one instance all proceeded. Ten per cent of payroll now goes on our above-the-line training investment.

SEMAP II next took shape as a critically important stage in the development of courageous managers was reached, and has been extensively documented[15]. Its focus was clearly stated as "enabling managerial growth and ownership succession". Using action learning, virtually every manager and supervisor and staff member in the enterprise was involved in either the senior set, or on two other levels known as A-MAP and B-MAP. Projects were chosen from across the enterprise on the initiative of the staff concerned, and worked on for six to nine months. And to the surprise of many – not least the owners – the major issues identified were information systems and logistics flows.

The production staff, once again, wished to move forward with further new technologies now available. They wanted to adopt a strategy for the capture of knowledge electronically that would enable their database of articles to be accessed in whatever manner customer demand was likely to ask in the coming decade. The investment, including training and transfer, was approved at around £500,000 on the advice of the staff concerned, who themselves called in the Printing Industry Research Association, PIRA, as their consultants in the process.

A Literati Club, being an authors' and editors' database, was proposed and implemented. A Promotional Logistics Information System was developed to improve the efficacy of the largest expense area in the enterprise. And a host more ideas were conceived. At six- and 12-month intervals after the major report-back workshops, the staff concerned were required to evaluate whether their proposals had been acted on, and if not why not – on direct report to the owners.

The significance of SEMAP II, with A-MAP and B-MAP, was, however, far greater than simply the projects conducted albeit that they were of major significance, with their implementation liquidating the investment in the programmes within a year. It translated into our next strategic thrust into systems development – coinciding as it went with the publicity surrounding the similar endeavours of Stayer[16] at Johnsonville Foods in the USA.

Systems strategy
The use of systems in our enterprise is, of course, as old as the enterprise itself. The difference between our systems strategy since the late 1980s and our previous approach is that we now develop and use them quite deliberately to make our managers more courageous. Previous systems had been either designed to increase productivities and efficiencies – such as the use of computer bureaux to process orders and despatches – or to provide intelligence to the owners concerning, for example, the cost effectiveness of promotional activities.

It was this latter area that first saw the awakenings of interest in systems that could enable our managers to learn incrementally[17]. Managers in the marketing services area, who today reinvest nearly 25 per cent of each year's gross revenue in the search for new sales either to existing customers or prospects, resolved to develop their own marketing database. None of the owners knew anything about such an activity professionally, although one with especial interest in information management immediately came to their assistance and as a sponsor. They organized their own search for an appropriate database package and downloaded the subscription/product-based lists from the organization's computer. Then they began to manipulate and work it to achieve their own purposes – clustering activities around a series of pre-agreed campaigns, and a campaign by campaign review cycle[18]. This non-owner initiative was the first occasion on which staff, unprompted, had led the owners to a better-managed business. The baton had begun to change hands and the owners' response was extremely positive. Spontaneous combustion had occurred. (We are not overlooking here the fact that the several action learning programmes had triggered all manner of management growth and indeed launched our highly successful acquisitions strategies; but the owners had led them out initially, and that is the distinctive point to be made.)

The marketing database was an almost immediate success – it took only a year to convince the Doubting Thomases among the managers' colleagues to make proper use of it. The owners willingly and dutifully gave it their strong seal of approval. However, its contribution to systems and strategic marketing in our enterprise was only at the beginning. The owners resolved to extend the modelling of information needs and flows across the entire organization, and to commit whatever resources were necessary to building systems that would enable our managers to run the enterprise.

While a holistic model was set down on paper for debate, at no time did the owners envisage anything other than a series of relational databases. And within such a framework, it was always to be, and still is, the intention that prototyping of systems away from the mainstream development would normally be encouraged. Mention has already been made how the SEMAP II programme had given rise to a promotional logistics information system and, of great importance for the future of the enterprise, a reshaped electronic production system. To these were added two customer clubs, and a club specifically for authors. Although a more modest investment than the reshaped electronic production system, the clubs were of very great importance for our courageous management.

The customer clubs covered the librarians and human resource professionals who were subscribers to our journals – populations of some 7,000 each. For each club, detailed profile information was collected and all relationships were channelled through specific teams of staff. These teams became our "experts" on the needs and wants and behaviours of those groups of customers. Rather than the owners having opinions or views – which of course they did – and using these as a basis for action, managers were now gathering their own intelligence which readily suggested what actions to take.

Librarian key customers purchase on behalf of academic colleagues across their institutions. To influence them and to serve them demands a different strategy from that required for human resource professionals – they buy for themselves, albeit with company funds. Their careers evolve and develop and as they do their needs for journals evolve. Librarians are much more likely to sustain subscriptions to a given journal for a longer period.

The authors' club, known eventually as Literati Club, had been proposed originally as a straightforward database to make the production task simpler to manage. It has gone forward to become one of the major engines for our current Quality Initiative in the enterprise, providing sample bases for benchmarking studies. It has become a routine resource for managers other than the owners (who it must be remembered are university academics) to grow in confidence in seeking good contributors, new editors, ideas for new launches and more besides. Most recently it has become the source of intelligence for the conduct of Author Development Workshops worldwide and for field sales visits at institutions where sales analyses show our products are under represented.

These clubs not only enable our managers to act courageously; they are now smarter and more intelligently briefed than the owners ever were[13]. It has been the success of these clubs in demonstrating how much more wisely well-designed information systems enable us to proceed that has most recently led to the enterprise's largest investment ever in systems development – our computer-based integrated customer environment (or ICE).

The conduct of the clubs using the marketing database download, and the conduct of the cross-selling campaigns, were monstrously tedious and inefficient. In comparison, the Literati Club's standalone system was jealously regarded. It was clear that sooner rather than later a new system for customer development would be needed. Its design was created from grass roots upwards and only to help implement it were consultants brought in. Everyone was asked for their wish lists, and a company-wide taskforce deliberated for a year or more.

At this juncture we went briefly off the rails. The customerization taskforce was a representative concern rather than a team of committed enthusiasts. Nothing happened until we got it back into the enthusiasts' hands to own. Then, even they made a false move too. Daunted at the size of the task of introducing an integrated customer environment embracing marketing needs, subscriptions control and financial collections, the enthusiasts invited in outside consultants who proposed to tell us what we needed. The owners intervened to say "no, thank you". The managers had to rise to the occasion no matter how daunting it might be! And if an extra in-company staff member was needed, so be it. The outcome, the ICE, had to be what its managers and operators specified and owned. And so it has been that the managers concerned have invested close on £1 million to get the ICE which they say they, not the owners, want and need.

Mentoring and coaching

We have never been wholly clear in our minds at MCB University Press whether the owners and increasingly, of course, the new senior managers as a group, needed to mentor and/or coach their colleagues. The two processes surely differ – mentoring is frequently defined as a listening activity and a questioning activity rather than the more prescriptive stance taken by a coach.

The owners and senior managers shared in an in-house mentoring workshop in 1990. As a result, we got a somewhat clearer understanding of what is involved. Rather than leading with coaching and then evolving to mentoring, we resolved to offer both behaviours at the same time – and, if forced to state a preference, to favour the deep end or stretching approach of mentoring. In other words we wanted individuals to do as much as they felt comfortable with, plus some stress, to the point where if they needed help they asked for it. Only if they got into strife was coaching to be thrust on them – as with the decision to return the ICE systems project back to the internal managers to lead.

Throughout the past decade, all in the emerging senior management team have had an owner/partner or company director as their mentor. It is that individual who each year conducts a self-driven appraisal and career review, and who also argues for the issues raised by the senior manager in the ultimate decision-making forum of the enterprise – the Board of Directors. Some of these dyadic relationships have worked well, some less well. They have at various times moved around by pressure of organizational role change rather than careful matching of the two involved. On their own, they could never have been sufficient for the task. The deliberate activities already described under management action learning that required senior managers to work at and share problems and issues – not least how they related to the owners – was vital too, and very effective.

Coaching in the main has come into play where Burdett[19] suggested it should. Managers used to managing "what is", needed guidance to adapt to "what can be or the art of the possible". Managers used to managing with formal authority were sometimes "reluctant to move into collaborative relationships as the basis for getting things done". The action-learning challenges demanded of all concerned created areas of dissatisfaction with the status quo while, at the same time, "affording a context in which that dissatisfaction could be addressed". For example, for several years there was far too little dissatisfaction for the owners' comfort in production. The SEMAP II programme transformed that and unleashed, just as SEMAP I before it had, an appetite for development that is now as hungry as that in the marketing and customer development areas. Indeed, production was first to come to terms with the major transformations of its work practices – most recently eliminating any own-warehouse activities in journal despatches.

Having created and focused on dissatisfaction, of course, the coaching and mentoring role seeks to ensure that it is suitably addressed. Each dyad will have found its own *modus operandi*. The secret seems to be not how proactive or reflective the mentor or coach may be, but whether the focus of the relationship is manager centred and intended to reach consensus[20].

Most recently, the senior managers themselves have launched what is called our Career Management Initiative. Twenty members of staff have personally requested that a senior manager should help them evolve and develop their career at MCB University Press. The allocation as between senior managers was made on their own initiatives and the owners were not involved. Once again we see an instance of senior managers showing the owners how to run the enterprise more effectively – and getting wholehearted support for their initiative. It is already showing good results after six months, with all clearly aware of what aspirations the 20 have. They are ready either to assist them to achieve them or to counsel why a particular ambition cannot be speedily achieved and what steps have to be taken as necessary preparation to get there eventually.

The owners have forced a further issue as well however. They have insisted that the senior managers take more responsibility for the "general" development and training of their colleagues over and beyond the focused skills needed for their current and immediately perceived roles. Management action learning programmes at MBA and Bachelor levels are again being offered to all levels of staff but this time under the aegis of those who have already learned their way to those awards. Known as our Enterprise School of Management, conducted in partnership with the International Management Centres, it has seen the development of a consortium approach with our enterprise's suppliers and of a new curriculum at Bachelor level, focusing on publishing in its broadest scope. The next generations of middle managers are not only being trained, mentored and coached by their senior colleagues; they are being "generally" educated under their leadership as well.

What we are seeking to create is an environment where the workplace is constantly challenged and stretched. We are seeking in every way to stimulate the people with whom we work without intimidating them. "Being a learning organization *per se* is not enough. The talents of each and every individual must be developed"[21].

Structural change
The formal structure of our enterprise was established in 1977 and remained unchanged until November 1993. The only element of doubt we regularly had about it was how best to attend to new launches and acquisitions. The reason we remained unchanged for so long was because we had been outstandingly commercially successful the way we were. The structure had been based on self-managed publishing/market missions – extremely fashionable then – in the heyday of output budgeting. For the ten previous years, foundation to 1977, it had been built around owner self-managed venture groups for each journal. So we have a not inconsiderable tradition of self-management in our enterprise.

The challenge to improve our structure emerged in high relief as the owners withdrew from their critical senior management roles as publishers, and invited those who had been their executive associates to take up the tasks. While the new incumbents had much of the knowledge required to do most of the job, it was inappropriate for the enterprise at large to ask it of them. The power and authority vested in an owner *per se* had given the role of publisher more than its nominal tasks, and that could not be transferred. Colleagues wanted to flatten the structure at senior levels as the owners moved upwards, just at it had been flattened by the owners previously at middle manager levels, not to create cardinals among themselves.

The first major fault lines emerged as previously indicated in the area of new launches and acquisitions. The publishers working as owners had always shared decision making in these matters as well as having responsibilities for their autonomous operational domains. First, an Acquisitions Unit and then a New Launches Programme were spun out of the orbit of the non-owner publishers, alongside the publishing/market missions, to be managed by the owners.

The second major fault line was more demanding. Future versus present issues could be handled comfortably in parallel, but a challenge to make the enterprise customer- as opposed to product-oriented could not. The emergence of the marketing database approach for promotional campaigns and development of existing customers was the death knell of the 1977 model. The marketing database was so evidently cost/beneficial that support for it grew with ever-increasing commitment to the need to give the whole enterprise a customer focus; the ultimate step indeed on this road was the creation of the ICE system. The organization had to be restructured as ICE came on stream.

What we have since come to realize was that, by following the logic of a customer focus, and establishing a new structure which none of the owners had ever operated in their time as senior managers, there was an opportunity on the grand scale for the new senior managers to call the shots. And they in their turn had the opportunity to do the same for their colleagues, who were taking up roles they themselves had never filled.

The need for the major structural change was perceived among the owners and leaders. It was promulgated in broad outline as VISION '94 in December 1992. It sought to align our structure with the future purposes of our enterprise[22,23]. Senior managers were then required to join in SEMAP III throughout the first nine months of 1993 to explore and test out the VISION for 1994. They were required to resolve how they would want to organize its implementation[24]. The top roles were offered by the owners to the key players in the former structure. Profit-related pay (PRP) was introduced for 1993 trading, and since continued, to ensure that so far as possible the enterprise did not take its eyes off the current year's bottom line while planning and thinking its way into 1994's new look. The only caveat entered was that at the outset the headcount was to remain constant, with any major gaps filled by contracting out.

At the same time, to ensure that the new structure and thinking got well beyond the senior managers, 20 staff members commenced action learning Bachelor programmes from across the enterprise, with their assignments and dissertations focused on areas of VISION '94 as debated with their senior managers. Three of the senior managers themselves joined an action learning MBA programme, again using their assignments and dissertations to focus on the development of the future in terms of empowerment, field sales campaigns, agent relationships, electronic publishing, facilities management and much more besides.

SEMAP III projects reflected the determination of VISION '94 to take a sharper look at our world markets – over half our revenue is from outside the UK. Schemes and actions emerged for greater penetration of selected continental European countries, and for the development of region-specific new launches and pricing strategies in Southern Africa and South America.

The integration of Publishing Logistics led to close focus on editorial procurement, a quality initiative for articles themselves, production tracking, journal despatches and copyright clearances. The determined customer focus of the new structure led to an evaluation of the quality of service offered in customer service – a strategy looking for "moments of truth" in the process, each to be better managed. The wholly new area of Sales Prospecting in contradistinction to Customer Development examined the key training requirements for that field – a precursor to the subsequent demands for a Bachelor programme based in publishing itself. Production similarly looked to ways in which its now arriving new technologies could enhance its performance, particularly by the development of middle management competences in the department.

Each of these projects can once again be seen to be at one and the same time creating and addressing dissatisfaction within the enterprise. The original restructuring that had taken acquisitions and new launches out of the publisher's domain was sustained but with a strong caveat. Now brought together in a Business Development Centre they were required to create no separate focus for developing the future. They were on the contrary required to take the initiative to bring together, across the organization, multifunctional teams to address the issues that were important, and then to get the outcomes carried into effect. For instance, as soon as an acquisition was mooted or a new launch explored, those concerned to see it was absorbed and/or implemented must be involved from the operational areas of the enterprise.

One year later, with VISION '94 firmly in place, two further dimensions have been added to this structural proactivity. The Head of Publishing Logistics has been asked outwith her operational responsibilities to initiate multifunctional teams to review and develop medium-term strategies for cohorts of journals within our total range, e.g. best sellers, best contributors, or acquired titles three years on. A new role leading out our Electronic Publishing Initiatives (EPIs), with the mandate to operate using the same patterns as the Business Development Centre, has also been established for a senior manager.

Our belief is that every individual in the enterprise must be as involved as possible in creating and sharing in the electronic publishing transformation of our product range and more besides in the coming years. The individual taking up the EPI role is one who has a distinguished track record as intrapreneur within the enterprise. As such there is no danger of losing the advantages of prototyping approaches to such activities which have enabled us to succeed well thus far. Nevertheless, we believe the time has come when the heart and substance of our existing technologies providing hardcopy publishing seem to be on the threshold of a discontinuity. We do not know what the outcomes will be when the situation stabilizes but we are ready to put ourselves out of one technology into another if that is what some or most of our customers require, and we can discern how to achieve it profitably. Everyone has a right to be involved in the excitement and it would be folly not to let everyone contribute as best they are able. As we observed in the opening paragraphs of this article we believe all our futures are at stake.

The success of the approach adopted during the implementation of VISION '94, i.e. promulgating a new vision and structure and inviting the senior management to explore how it wanted to implement it, has staggered us all. It will be clear from our analysis that by almost any definition our enterprise has enjoyed a very considerable pattern of empowerment, been a learning organization and used strategic management development over the past decade. But this was a totally new experience for the owners and senior managers alike. "The battleship", as Carr[25] describes it, "had become a fleet".

Resistance to change? Why? By the end of 1993 the staff simply could not wait to get started. And that was before the fruits of the ICE system were going to be available – not ready until August 1994. The changeover was so smooth it was hard to credit that what had gone out of the window was the way of organizing ourselves that had built the enterprise to what it had become – the structure that had enabled the owners to build a truly international academic management publishing house. Apart from the vision formulation and a determination to set in place the programmes of management action learning, the owners had transferred responsibility to the senior managers. They had created an environment in which managers had been keen and able to identify and solve complex problems on their own. The managers concerned chose to create power for themselves. As Harari[26] reported from another educational publishing company, any owner could say: "I don't know what my people are doing but, because I work face to face with them as coach, I know that whatever it is, they are doing exactly what I'd want them to do if I knew what they were doing!"

By June 1994, for the owners' Think Tank sessions, the senior managers were able to report no major interface problems between their new areas of responsibility in the new structure. What they did not say, and what the owners have readily conceded gives them very great pride, is that a host of temporary teams and taskforces have been spawned, worked matters through and disbanded over the past 18 months[27].

A host of individuals, at all levels, have called together colleagues from across the enterprise in a lattice of relationships to address issues from the design of reports to the creation of a new sample copy strategy or the relaunch of an ailing journal[28-30].

They had little hesitation in so doing because they could see no real barrier to doing it and it was the obvious thing to do to solve their problem. Nobody had any ready answers to offer because it was all going to be different.

Leading the courageous on
As we observed at the outset, we have some bad news, some threats to overcome. In the context of what has been described, of what the owners have done and led us to do over the past decade, we believe we are reasonably well

equipped to evolve a better scenario than selling off the enterprise to other owners. We believe we will be able to address the challenges of electronic publishing in a profitable way. With a continuing coaching and mentoring role from the owners and our permanent commitment to action-learning development, we have reached the stage where we can sustain our systems and structural growth and development.

Notes and references

1. Weis, P., "Achieving zero-defect service through self-directed teams", *Journal of Systems Management,* February, 1992, pp. 26-36, whom we thank for suggesting the notion of "courageous managers".
2. Simon, D., "Managing cultural change at BP", *Target: Management Development Review,* Vol. 4 No. 3, 1991, pp. 16-19.
3. Andrews, D., "The trust factor: the hidden obstacle to empowerment", *APICS – The Performance Advantage,* May 1994, pp. 30-2.
4. Temple, R.E. and Droege, R.W., "Internal customers need delighting too", *Managing Service Quality,* Vol. 4 No. 1, 1994, pp. 14-17.
5. Hand, M., "Freeing the victims", *Total Quality Management,* Vol. 5 No. 3, June 1993, pp. 11-14.
6. Kotter, J., *The General Managers,* Free Press, New York, NY, 1982.
7. Ripley, R.E. and Ripley, M.J., "Empowerment – the cornerstone of quality: empowering management in innovative organizations in the 1990s", *Management Decision,* Vol. 30 No. 4, 1992, pp. 23-43.
8. Revans, R., *The Origins and Growth of Action Learning,* Chartwell Bratt, London, 1982.
9. Peters, J.V., "Customers first – the independent answer", *Business Education,* Vol. 9 No. 3/14, 1988.
10. Wills, G., *Your Enterprise School of Management,* Revans' Professorship Report: 1992, MCB University Press for International Management Centres, UK, 1993.
11. Sutton, D., "The problems of developing managers in the small firm", *Training and Management Development News,* Vol. 1 No. 1, 1987.
12. McClelland, S., "Gaining competitive advantage through strategic management development", *Journal of Management Development,* Vol. 13 No. 5, 1994, pp. 4-13.
13. Bruce, B., Jordan, T. and Wills, G., "Realizing the benefits of a marketing intelligentsia", *Marketing Intelligence & Planning,* Vol. 12 No. 6, 1994, pp. 21-34 (see Chapter 14).
14. Newton, R.J. and Wilkinson, M.J., "Project morale: the empowerment of managers in their everyday work", *Empowerment in Organizations,* Vol. 2 No. 1, 1994, pp. 25-30, explore a similar approach using action learning at Ashworth Hospital.
15. Wills, G., "Enabling managerial growth and ownership succession", *Management Decision,* Vol. 30 No. 1, 1992.
16. Stayer, R., "How I learned to let my workers lead", *Harvard Business Review,* November/December, 1990.
17. Argyris, C., "Double loop learning in organisations", *Harvard Business Review,* September/October 1977, pp. 115-25.
18. Wills, G., Bruce, B. and Duncan, T., "Creating a marketing intelligentsia", *Marketing Intelligence & Planning,* Vol. 9 No. 4, 1991, pp. 1-20 (see Chapter 1).
19. Burdett, J., "To coach or not to coach? that is the question", Parts 1 & 2, *Industrial and Commercial Training,* Vol. 23 No. 5, 1991, pp. 10-16 and No. 6, 1991, pp. 17-23.

20. Mumford, A., *Management Development Strategies for Action,* Institute of Personnel Management, UK, 1989.

21. Casse, D., "People are not resources", *Journal of European Industrial Training*, Vol. 18 No. 5, 1994, pp. 23-6.

22. Harker, W.C., "Alignment for success in the nineties", *Business Quarterly*, Winter 1991, pp. 107-12.

23. Chorn, N.H., "Organisations: a new paradigm", *Management Decision*, Vol. 29 No. 4, 1991, pp. 8-11.

24. Carr, C., "Managing self-managed workers", *Training and Development*, September 1991, pp. 37-42.

25. Carr, C., "Empowering leaders", *Training and Development,* March 1994, pp. 39-44.

26. Harari, O., "Stop empowering people", *Small Business Reports*, March 1994, pp. 53-5.

27. Lynch, R.F. and Werner, T.J., "A league of their own", *Small Business Reports*, April 1994, pp. 25-42.

28. Shipper, F. and Manz, C.C., "An alternative road to empowerment", *Organizational Dynamics,* Vol. 20 No. 3, Winter 1992, pp. 48-61.

29. Martin, L. and Vogt, J.F., "No sense of trespass: empowerment through informational and interpersonal licence", *Organizational Development Journal,* Vol. 10 No. 1, 1992, pp. 1-8.

30. Belasco, J., "Empowerment as a growth strategy", *Management International Review*, Vol. 32 No. 2, 1992, pp. 181-8.

Chapter 8

ROI in management development

It is widely believed, and we are among the believers, that management development programmes are a good investment. We believe they are good for the people who participate. We believe we know from regular experience that they are good for the people who stay behind and who, by default, then get more scope to flex their muscles and brains when the favoured one is away. But there are seemingly no analyses where such beliefs are systematically checked out with data of the hard programme return on investment (ROI), at the end, then one year and five years later.

Between 1992 and 1995, we tracked what occurred when over £3 million was invested on MBA programme fees to develop 300+ managers, in 12 countries around the world. Questionnaires were completed and their contents debated for six hours at graduation time which was held consecutively in Kuala Lumpur, Surfers Paradise, at Ripley Castle in Yorkshire and Amsterdam. The MBA programmes were organized by International Management Centres (IMC), the leading action learning multinational business school following the principles set down by Reg Revans over 50 years ago.

This study ($n = 101$) covers just one part of what has now, over 12 years, become a £30 million tranche of investment made for over 5,000 managers by nearly 1,000 enterprises in 31 countries using the processes of action learning with IMC.

Best estimates indicate that the 300+ managers spent 2,500 hours each, 750,000 hours in total, talking with one another, with colleagues, sharing their managerial problems and challenges and helping one another to come to terms with actions which needed to be taken. IMC faculty members, skilled and knowledgeable in their own areas, participated in their meetings. Their role was to help the managers to be aware of what was already known in the areas where they were seeking to improve and to help them to discover new knowledge where necessary. The 5,000 managers in the overall programme spent nearly ten million hours at the task.

Just under half the 300+ managers went on to implement most or all of their recommended courses of action with the support of their sponsoring organizations. These typically required investments which were well in excess of five times the programme fees and on not infrequent occasions in excess of £1 million. These investments in their turn yielded satisfactory returns, through major cost reductions and new revenues.

My employer gave me full opportunity, responsibility and freedom to implement my projects.

With Carol Oliver,
Management Development Review,
Vol. 9 No. 1, 1996

My employer invested £1.2 million to fully implement the project. We gained a threefold return in the first twelve months.

In addition, of course, a wide range of non-financial benefits were cited. This analysis suggests that the 300+ managers triggered at least £10 million of investment to implement their action plans, with ROI expected of £50 million. Judgemental extrapolation suggests the 5,000 probably triggered over £100 million with half a billion pounds ROI. Certainly some programmes!

This particular quantum of managerial development by action learning, *par excellence,* exemplifies the case for investing an enterprise's funds in such programmes regardless, and we say regardless advisedly, of whether the manager now with an MBA qualification stays with or leaves the sponsoring enterprise. And regardless of how much the individual *per se* may or may not have learned. Which is not intended to belittle individual learning for a single moment. Our purpose here is to argue that soft benefits do not need to be educed to justify management development of the action learning variety. More and more of it should be done within any and every enterprise until the marginal hard ROI equates with other projects in hand. Educationists should cease being afraid of measuring and propagandizing the hard benefits they provide to enterprises.

Did the managers leave?
Well no. Of course not. Most of them stayed put, relishing their new intellectual understanding and involvement in their enterprise. Would you leave an enterprise that has sponsored you on an MBA programme that gave you the broader helicopter view of the business way beyond your previous experience, enhanced your self-confidence, presentation, listening, team and overall people skills, given you understanding in particular of financial issues, helped you to become more action-oriented on the critical issues affecting the enterprise, and more timeous and objective in the conduct of your role? That is exactly what they reported had happened:

> We found the best solutions often came from those closest to the job. It was dangerous to have our own preconceived ideas or to think we knew best.

> As an action learner I was able to gather information more easily than as a manager; others wanted to help me – even commiserated with me.

> We gained confidence to talk to people at all levels in the enterprise.

> In developing our projects we learned what the other parts of the company did, their purpose and methods of working.

All this, it can be seen, was accomplished in the context of the organization itself, and their dissertation project (while academically rigorous) had to be based on the key ingredients of the future strategies of the enterprise. The conclusions reached normally made their way directly to the boardroom. IMC invariably involved top managers in the process as mentors and sometimes as adjunct faculty. IMC ensured that bosses were fully aware at all stages of the action learning approach. Subordinates, colleagues and families who took up some of the manager's load while diverting 2,500 hours to the programme were, by and large, grudgingly supportive too.

Ninety-three per cent of managers were still with their employer at the end, and 48 per cent had been promoted during the 24-month programme.

The MBA dissertation project was seen as the runaway winner among all the assignments in terms of personal and organizational benefit. It also took the most sweat and tears to create. After that, each of the specialist areas of management intelligence, corporate integration, human resource management, finance, marketing and operations had its own following. These preferences were reassuringly based first on usefulness to the enterprise and eye-opening potential, and only latterly on the faculty member's particular style and approach to supporting the learning processes. Very few were dismissive of any areas, but where there were regrets it was that the debate had not been vigorous enough.

Which learning resources were most valued?
All manner of resources are available to a work-based learner. Yet the greatest contribution of all came from what Revans describes as "comrades in adversity". These are the fellow members of the small learning cell that action learning uses, known as the Set. On a semantic differential scale, it averaged 3.90 with a maximum score of 5 available:

> Feedback from the Set, the level of interaction and exchange of ideas were both fun and enlivening – as well as invaluable.

The second greatest contribution came not from the subject area experts but the Set Adviser, or linker/facilitator with a score of 3.82, which was followed by the tutors with the expertise at 3.75. Next came the course materials issued and found in support (3.66), colleagues at work (3.18), bosses and mentors (3.06) and finally library services with 2.83.

One of the most valued but underestimated resources which managers use is feedback from tutors through the marked assignments that go towards the ultimate award of the MBA. It is, of course, a crucial moment for faculty to make sure the measures of quality really do justice to the context in which the manager concerned works and seeks to act and improve performance. With programmes in 31 countries over 12 years ranging from South Pacific islands and African homelands to the more mature Dutch, UK or Australian economies it is no small challenge. The verdict given by two-thirds of the managers was that the process was "good or very good":

> After having done all nine assignments, the habit developed of going through the four learning stages is continuously with me.

> Receiving my first A grade for an assignment was the greatest highlight.

> I struggled with my Organisational Management assignment, but when I got a good grade it gave me confidence, it was a turning point. I knew I could do it.

How typical were these managers?
There are few ways in which we can assert that the managers who devoted 2,500 hours each to these IMC programmes to gain the MBA are normal. Normal managers simply do not do this sort of thing. They do not seek and relish the intellectualization of their role as a manager. They live off their wits and their experiences. And why not?

Action learners believe that wits and experiences can be very considerably improved on when they are shared in a rigorous and disciplined way.

Yet it would be erroneous to believe that IMC's managers were a scholarly segment, although 82 per cent did indeed have Bachelor level university education (44 per cent) or its equivalent in a professional area (38 per cent). Eighteen per cent were straight from their experience – often performing outstandingly on the programme as well as in their workplace.

At the end of the programme all were adamant that they had, for the first time in their lives, learned what learning was. It was continuous, of course, but most excitingly, they had learned that all managers, all team members, have potentially got different styles and preferences for learning and for working. Unless each manager knows, understands and accommodates these differences the workplace cannot hope to perform to its best. In this respect they believed they were henceforth and irrevocably a different sort of manager.

What did we all do wrong?
If the whole action learning project sounds like an outstanding success, that is simply because it has been. Our weaknesses at IMC were mainly associated with faculty members who found it difficult to meet the expectations of demanding, action and context-based managers. The second criticism was at the place of work, with top managers and colleagues giving less support than was hoped for. The third complaint was of IMC's organizing skills for Sets around the globe. And finally, managers looked at themselves and criticized their own inabilities to manage their time effectively.

Our response to each of these, which have been recurring themes for the whole 12 years, has been to empower managers to hire and fire their faculty, to provide boss mentoring development workshops whenever possible, to evaluate and re-engineer our own support organization (most recently leading us on to the Internet at URL http://www/mcb.co.uk), and to introduce time management tutorial workshops at the beginning of all our programmes. These four themes of criticism seem endemic to the process itself and accordingly set to remain with us for ever. Forewarned thus, we have a continuing search to alleviate these complaints permanently on our managerial agenda.

One year and five years later
Since 1982 it has been a requirement of all IMC graduates that they must renew their professional competences at least every five years and attest to how they have achieved it. It is known as the Five Year Continuing Renewal. Experience has shown that five years after a major action learning project is normally too

late to seek to evaluate its specific benefits. Events will have moved on and almost always career promotion will have removed the individual concerned within or beyond the enterprise which sponsored participation on the programme.

After one year, however, where we have introduced a progress audit option for graduates known as A+, we have had much greater success in seeing the extent to which their end-of-programme assessments have held up. They afford the supporting evidence we need in the round, albeit with pluses and minuses present. For some, the hoped-for implementation did not happen with: "the organization is now focusing on alternative structures and the original idea has been discounted". For others, the outcomes were better than expected: "They actually implemented".

Conclusion

Our conclusion must accordingly be that while soft benefits will continue to deliver well after the programme is over, through changed behaviours, growth in confidence and the like, most of the sensibly attributable hard ROI is realistically traceable by thorough, evaluative survey methods on completion and during the 12 months following a programme.

Training and development managers should, in our opinion, do a great deal more evaluative research to measure the organizational and particularly the financial impact of programmes. In this way, a budgetarily supportive culture can emerge to supplement the generalized feeling of goodwill towards development of staffs.

Further reading

"2nd Quinquennial Review of MBA: November 1992", *Design & Process Newsletter,* International Management Centres, Buckingham, 1992.

Barker, J., "1991 David Sutton Fellowship Report", *Design & Process Newsletter,* International Management Centres, Buckingham, 1991.

Bennett, R., "Effective set advising in action learning", *Journal of European Industrial Training,* Vol. 14 No. 7, 1990.

Caie, B., "Learning in style – reflections on an action learning MBA programme", *Business Education,* Vol. 9 No. 3/4, 1988.

Coates, J., "An action learning approach to performance review and development", *Business Education,* Vol. 9 No. 3/4, 1988.

Cusins, P., "Action learning revisited", *Industrial and Commercial Training,* Vol. 27 No. 4, 1995.

Espey, J. and Batchelor, P., "Management by degrees: a case study in management development", *Business Education,* Vol. 9 No. 3/4, 1988.

Gore, L., Toledano, K. and Wills, G., "Leading courageous managers on", *Empowerment in Organizations,* Vol. 2 No. 3, 1994 (see Chapter 7).

Margerison, C., "1991 Revans Professorship Report", *Design & Process Newsletter,* International Management Centres, Buckingham, 1991.

Mumford, A., "Developing managers for the board", *Business Education,* Vol. 9 No. 3/4, 1988.

Mumford, A., "Effectiveness in management development", *Business Education,* Vol. 9 No. 3/4, 1988.

Mumford, A., "Learning in action", *Personnel Management,* Vol. 23 No. 7, 1991.

Mumford, A., "A review of action learning literature", *Management Bibliographies & Reviews,* Vol. 20 Nos 6/7, 1994.

Mumford, A. and Honey, P., "Developing skills for matrix management", *Business Education,* Vol. 9 No. 3/4, 1988.

Peters, J., "The new MBA – what it means for managers", *Business Education*, Vol. 9 No. 3/4, 1988.

Prideaux, G., "Pan setting II", *Training & Management Development Methods*, Vol. 5, 1991.

Prideaux, G., "Making action learning more effective", *Training & Management Development Methods*, Vol. 6, 1992.

Revans, R., *The Origins and Growth of Action Learning,* Chartwell Bratt, Bromley.

Revans, R., "The learning equation: an introduction", *Business Education*, Vol. 9 No. 3/4, 1988.

Seekings, D., talks to Brian Wilson, "Allied Irish Bank in Britain: organizational and business development through action learning", *Business Education,* Vol. 9 No. 3/4, 1988.

Smith, A., "1992 David Sutton Fellowship Report", *Design & Process Newsletter,* International Management Centres, Buckingham, 1992.

Sutton, D., "The problems of developing managers in the small firm", *Business Education*, Vol. 9 No. 3/4, 1988.

Sutton, D., "Action learning in search of P", *Industrial and Commercial Training*, Vol. 2 No. 1, 1992.

Thomas, J., "Researching learning to learn", *Design & Process Newsletter,* International Management Centres, Buckingham.

"Total Quality Assurance", IMC's 4th Annual Professional Congress, Buckingham, England, November 1992.

Wills, G., "A radical alternative in management education", *Business Education*, Vol. 9 Nos 3/4, 1988.

Wills, G., "The customer first – faculty last approach to excellence", *Business Education,* Vol. 9 Nos. 3/4, 1988.

Wills, G., "Wealth creation through management development", *Business Education,* Vol. 9 No. 3/4, 1988.

Wills, G., "Action learning pan setting", *Training & Management Development Methods,* Vol. 5, 1991.

Wills, G., "Managing networking", *European Journal of Marketing,* Vol. 25 No. 4, 1991.

Wills, G., "1992 Revans Professorship Report", Enterprise School of Management, *Design & Process Newsletter*, International Management Centres, Buckingham, 1992.

Wills, G., *Your Enterprise School of Management*, MCB University Press, Bradford, 1993.

Wills, G. and Day, A., "Marketing and selling at work: the IMCB/NatWest Management Development Programme", *Business Education*, Vol. 9 Nos 3/4, 1988.

Zuber-Skerritt, O. and Howell, F., "Evaluation of MBA and doctoral programs conducted in Pacific Region", report submitted to the International Management Centres, Pacific Region, 1993.

Appendix. The value of career development courses

Case study: the St Helier NHS Trust MBA programme
In August 1995, shortly before the completion of its first MBA programme, members of the set were interviewed at length to investigate and identify their perceived return on the investment by the trust, in both qualitative and quantitative terms. The interviewees had either completed or were in the process of finalizing their dissertations. The results, even at that early stage, indicated a total estimated annual efficiency saving in excess of £2 million, with numerous benefits by way of quality improvements in services and the managerial skills of all set members. Less tangible,

but no less important, were the positive perceptual and cultural changes affecting set members and their colleagues.

Measuring benefits

In a great many of the projects, improvement in quality of service to clients and customers was of paramount importance. Savings, though frequently substantial, were not the prime motivator but the by-product of increased efficiency and effectiveness.

Quality benefits were often recognized as immeasurable; for example, improvements in communication and understanding between individuals and departments or a total change in the culture of a service unit.

The cost of failure concept was introduced as a member when one associate student, unable to put a monetary value on work that would have been done anyway, recognized that failure to deliver an IT project successfully and on time, could lose the trust £1 million in Department of Health funding. The role of the MBA in facilitating successful delivery of this project was estimated as an increase from 80 per cent to 99 per cent in confidence that the project would be delivered successfully and on time.

Quality benefits relating directly to the service provided to clients and customers were perceived by most associates to be critical to the trust's success, inasmuch as excellent service generates more contracts, referrals and income, whereas poor service could affect the trust's reputation, losing goodwill and custom not just from the offending service but across the trust as a whole. With five other trusts close by, all associates seemed aware of the importance of winning and keeping clients. Several had already been successful in wresting contracts from other trusts, private services and public bodies.

Because of the independence of service units, it was recognized that work done in an associate's own area, or in another area for a particular project, had implications trust-wide.

Three associates stated that all their MBA work had been done in their own time. This was estimated as at least £20,000 each of unpaid effort on projects.

Occasionally, own time spent was converted into management consultancy hours which could be costed at upwards of £1,000 per day, but it was generally recognized that the trust would have been likely to call a management consultant in on that particular project. Those associates with experience of management consultants were spontaneously negative about their efforts and put a high value on projects being done by people who understood the context and the issues.

Since the nature of an action learning MBA is to deal with real issues in the workplace, it is inevitable that, sooner or later, most issues would have been addressed by someone. It was, therefore, important to try and assess what role the MBA had played in the successful completion of projects.

Overall, there were several distinct categories of observation.

Time

The MBA got the project started, or finished, much sooner than it might otherwise have been by, in some instances, up to two or three years. This meant that benefits could be felt much earlier by the trust.

Organization

The MBA meant that projects were tackled in a more organized, vigorous fashion with less trial and error, time wasting experimentation and fewer mistakes.

Communication

Improved communication, upwards, downwards and sideways between individuals and departments inside and outside the trust, was continually cited as a direct benefit of the MBA programme and one that would be of great and lasting value to the trust.

Confidence
Associates claimed to have gained confidence to ask questions, try things out, push things through, tackle difficult issues, take risks (with a successful outcome), discuss matters with other departments and senior staff.

Appropriateness
Several associates felt that their understanding of the issues, culture and professional concerns of those involved brought substantial advantages that would have been absent had they been tackled by less experienced staff, by management consultants or by working parties, which might have happened without the opportunities the MBA created.

Culture
Several associates experienced significant changes in the culture of their service units. Mostly these were attitudinal and related to either the creation and/or empowerment of teams or the move from being reactive to proactive when it comes to strategic issues.

Footnote
One associate was convinced that, whatever the value of the immediate benefits of MBA 1, the true benefit to the trust would show only in 18 months to two years time, as the associates developed their services.

Chapter 9

Networking and its leadership processes

"Networking" is a term now widely used to describe two contemporary organizational empowerment phenomena, both superficially the same but, in fact, fundamentally divergent.

The first phenomenon is a deliberate extension of matrix management within larger organizations, that gives increased importance to cross-functional work groups focusing on corporate outputs (Feneuille, 1990; Huey, 1994). Such networks are empowered to energize the sclerotic enterprise and, in many cases, have achieved the seemingly impossible. They are the sociological equivalent of the intrapreneurship movement launched by Pinchot (1985). Huey (1994) calls it the "age of post-heroic leadership".

The second phenomenon does not originate within large organizations at all. It is a coalition of separate individuals voluntarily working together to achieve a common purpose or goal and who, in order to achieve that, are prepared to empower some among themselves to act as leaders and catalysts in the best interests of the network (Snow *et al.*, 1992).

The similarity between the two phenomena lies in their rejection of hierarchical and functional models of organization, based on authority and specialist expertise respectively. In place of these models "networking" places the market as the engine of organizational purpose (Ghoshal and Bartlett, 1990; Theuerkauf, 1991). The marketplace and the customers within it who have transactions with the network determine who shall be involved for any particular purpose and how the relationship shall be focused to be conducted effectively.

The fundamental difference between the two phenomena is that socio-intrapreneurship seeks to make the large organization work better to help it to regenerate itself, while the socio-entrepreneur has no wish to live or work in a large enterprise at all. The networked socio-entrepreneurs are combined to achieve a democratically shared purpose. They also accept that it is their own responsibility to disengage from the network whenever its purpose can no longer be personally shared either temporarily or permanently.

The difference can be clarified using Pinchot's most memorable Law of Intrapreneurship – "Come to work every day willing to be fired". It implies that you have an employer who has your fate in his hands. The networked individual reserves the right to hire and fire to himself alone.

This article will review what has been learned and published lately about the realities of networking both among socio-intrapreneurs and socio-entrepreneurs. But, most importantly, it will review the processes of leadership involved – be they delegated downwards by fits of empowerment in the large

Leadership and Organization
Development, Vol. 15 No. 7, 1994

organization or serendipitously created in the democratic framework of the network of socio-entrepreneurs.

Such purposes are strictly selfish to the faculty and graduate members of International Management Centres, a network of socio-entrepreneurs. They must currently exercise their responsibilities as electors of their second generation of leaders. And it is ironic but not, perhaps, paradoxical that the task of undertaking this review should have fallen to myself as a first-generation leader. My first generation has refused to find its successors. We have challenged the network to decide for itself who might best lead them within a framework of their shared goals. But I have personally accepted their return challenge to draft a job specification and an induction guide for our successors. They assert that the first generation's insights linked to an evaluation of published literature is potentially highly valuable not only for their own purposes but also for networks more widely. The literature is replete with success stories, often journalistically presented. There is a dearth of insider critical analysis of the downside problems, or balanced advice on the issues.

The current nature of IMC as a network
The origins and evolutionary growth of International Management Centres (IMC) as a network have been described in detail already (Wills, 1992). Since 1992 we have progressed from a coerced network, through co-ordination to a nascent co-operative (Johannisson, 1987).

IMC is now a socio-entrepreneurial network with the shared goal of helping managers to learn how to be more effective and to act on that learning. Below is the statement widely communicated within the network.

IMC's philosophy of management action learning
IMC aims to assist in the development of effective managers by designing a learning process which is both intellectually stretching and founded on the real problems and opportunities facing managers in their own environment. We call it action learning.

The key conceptual and practical framework for this is provided through:

- A diagnosis of the issues on which managers have to be effective in their work environment and the necessary competences.
- A learning partnership between IMC and the client organization.
- Channelling developmental processes through taking action on real problems.
- Positive help to individuals on their personal learning processes to aid current and continued development.

The major outcomes of IMC's approach are:

- Improved effectiveness of managers in their current and future jobs and thereby in their personal careers.

- Learning benefits to individual organizations and eventually to national economies through productive work on organizational objectives, opportunities and problems.

- A continuous contribution to the body of knowledge and the identification of desirable actions on the effective combination of real work and managerial learning.

To achieve that shared goal IMC has evolved design, marketing, delivery and evaluation programmes that are based on an educational philosophy known as action learning. In line with the market and customer focus of networks IMC requires its students (known as Associates) to specify their curriculum issues that shall be the vehicles for learning. Once these are manifest, the appropriate individuals within the network to achieve the goal are brought together as the tutorial team.

Such an approach is well distanced from the traditional hierarchical or functional models of management education and development which assert that the faculty know best as experts, and as ones in authority. Students are there to learn what the academy believes should be learned.

Without the creation of a network approach, action learning programmes cannot be delivered, market focus cannot be the goal. The concept is perhaps closest to relationship marketing (Blankenburg and Johannson, 1992).

Any network member making the sale of an action learning programme knows he can call on other members to assist in its delivery and evaluation. To make the sale in the first instance, he will have relied on network members to assist in the design and marketing activities. This neatly, but oversimplistically, dichotomizes the leadership processes in IMC. The programmes have first to be sold and second, tutored. The oversimplification arises because a sale can take place only if the customer has very reasonable expectations of effective delivery, and the tutorial can be efficiently accomplished only if the right programme has been designed and marketed in the first place. As such, goal achievement for the network depends greatly on how the two elements support and strengthen one another rather than pull apart, notwithstanding the clear truth that those who most powerfully deliver and evaluate programmes are seldom those with the skills or understanding to design or market them to the same high level, and vice versa.

Such potential for misunderstanding and failure to appreciate disparate skills and their vital contribution to effective functioning of the network is exacerbated when IMC delivers programmes for awards at Bachelor, Master and Doctoral levels. Evaluation in those particular circumstances is, of necessity, outside the formalized network in the hands of "external" examiners. "Standards" that are all too frequently of the functional, not market-focused, variety are applied.

The IMC has one further dimension, which is the inescapable outcome of its adoption of action learning as its educational philosophy. It designs, markets, delivers and evaluates its programmes wherever the market might wish. And over the past 12 years that has meant 31 different countries from Finland to New Zealand, Hong Kong to Argentina – in four different languages.

Any suitably committed entrepreneur is readily welcomed into IMC's socio-entrepreneurial network, provided that he is willing to be appropriately inducted and to abide by the network's rule book as it exists from time to time. This network discipline, obviously of vital significance to the cohesion of the network, is not intended to be oppressive. As soon as a new member has been inducted he becomes a fully participating member in the formulation of all future policies and practices, which are incorporated in IMC's rule book, known as the Conspectus. All network officers are specifically elected by one or other of the governing bodies – the Council or the Common Multinational Academic Board.

Why networking now?
There seems to be considerable unanimity in the literature on networking that it is the next or coming pattern of organization for most if not all of us at work (Watson, 1990). It is argued that economies of scale can no longer outweigh the fleetness of foot that a network affords. The network is a hypersensitive ecosystem, responding by shedding or taking on resources on an almost daily basis without the baggage of employment contracts, or politicization of decision making. New skills can be accrued almost overnight to tackle whatever challenges might arise.

Not only does the speed of change in the competitive and technological environments of any enterprise make flexible responses an imperative for success, however. The human resource movement argues most convincingly that the better and better-educated individuals seeking employment are unwilling to subordinate themselves to anything less than a work activity that engages their intellect. After all, most non-intellectual activities can be and are being automated in our current era.

However, writing from the inside of a socio-entrepreneurial network that has evolved a very long way since its original coerced launch in 1982, the downside issues are very tough indeed. All networks have to hold their members together by vision and shared purpose – a familiar enough observation in an hierarchical or traditional enterprise. The greater significance of the challenge of vision and shared purpose in a network is that, unless it is very strongly projected, members simply melt away. They do not continue to come to an office every day. To get the vision and shared purpose up and running in the first place, and to sustain it in a coherent fashion, is as improbable a goal as it has been discerned to be for Peters and Waterman's original "excellent" companies.

Continuous renewal is the inescapable requirement in a network (Theuerkauf, 1991). This requires, as elsewhere, a willingness to let good ideas occur anywhere at any level. The network leader must see his role as facilitating any and every one to be able to work their ideas into the network, searching for kindred spirits who want to share their pursuit.

Charan (1991) describes this restlessness as the necessary social architecture of a network. A robust network does not imply harmony among members. More frequently it will be characterized by heated debate and legitimate disagreements. Network leaders must see their responsibilities as to encourage such ferment, and they must have the strength to get the network through it and out the far side to a renewed understanding of shared purpose.

Enthusiasm for the network at any time will be a measure of the extent to which its members feel an intense benefit from belonging and value the quality of interaction they achieve when they come together. This does not mean that networks need to reach agreement on every aspect of their vision or goals. But it does mean that they need the highest professional alignment on specific tasks.

At IMC the intensity of the debate and argument is clearly noticeable – particularly during the first working sessions of each major get-together. It all gushes out – the new ideas, the criticisms, the injustices. Only when these have been addressed, if not resolved, can the network settle down to mature debate and find its common ground for the next phase of its activities.

We have learnt from ancient times that democratic institutions are well lubricated by bread and circuses. For us the bread takes the form of a shared understanding of how each can be a more effective tutor or salesman of action learning programmes. The circuses are our Annual Multinational Congregations and local Admissions Ceremonies, accompanied by robes and formal certificates.

While our multinational spread of activities has made regular, sustained communication almost impossible until recently, we have used the circus magnetism of our 12 annual Ceremonial Congregations and, for six years now, mid-year Annual Professional Congresses to encourage, even structure, a plethora of fringe meetings.

Facilitating multinational communications

I say "until recently" because, as a result of upward pressure from Australian and Dutch members, IMC has pioneered the first multinational two-level Bulletin Board System (BBS) in any academic institution. Our electronic Atherton Intelligence System (AIS), described in Figure 1, enables any member of the network to access any other member, anywhere in the world. It has transformed the up-to-dateness of all resources offered to faculty, graduates or associates worldwide, and gives the opportunity for effective upward suggestions for network development for the first time.

Such seemingly trivial matters as up-to-date address and telephone/fax numbers, which are an eternal problem for any multinational organization, cease to be such problems. Data's integrity improves as well as its quality and timeliness. But most importantly of all, the centre loses all perceived need to coerce. Our electronic BBS truly makes our network multinational in substance as well as name and its leaders have the wherewithal to listen and learn. From that learning they can discern how to orchestrate regeneration and with whom, or in what coalition.

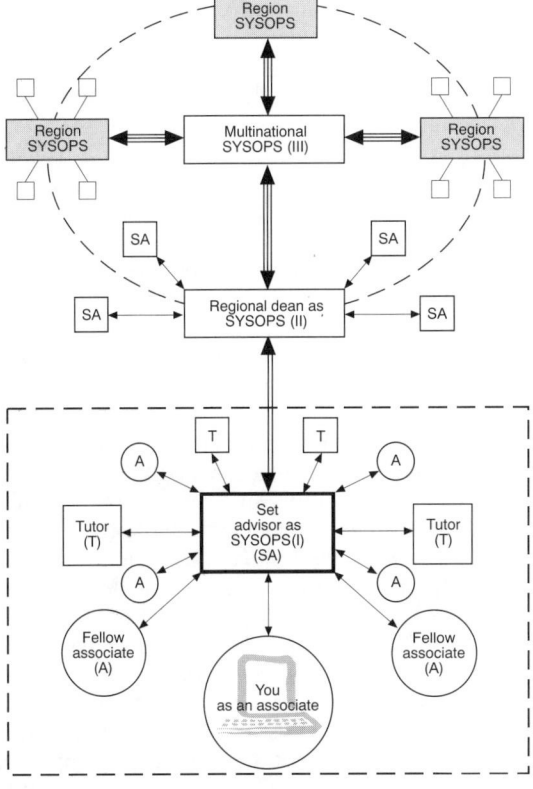

After Graduation

Everyone who has worked in a set using AIS has the opportunity to continue working with all fellow set members and tutors through AIS for ever and a day. The software is in place, and fully paid for after all. There are at least three reasons why we must do just that!:

1. As a graduate member our membership services will flow through the AIS. Our ANBAR updates and our full text retrieval are triggered within AIS. Our new members' *Electronic Newsletter* and information notes all come from the AIS – including our welcome subscription requests, reminders and bad debtors' pillories!

2. During the 12 months immediately following graduation, our *A+ enhancement* of our award is chased and attended to through AIS. Annually IMC asks members to notify their ongoing research activities and to give details and judgements on their continuing management development and career achievements and learning. Specifically *after five years*, IMC expects all graduate members to evaluate all this to qualify for continuing membership and, if wished, companion membership as well.

3. But over and above, before and after, the two points above, most sets want to stay together, to socialize together and much more. The AIS messaging service is still there, as it was througout the programme, for whatever use is convenient.

Key:

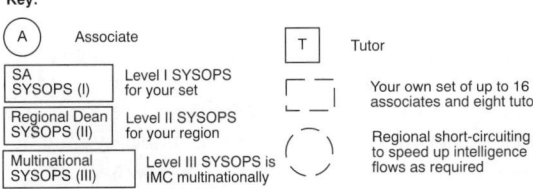

(A) Associate	T Tutor
SA / SYSOPS (I) — Level I SYSOPS for your set	Your own set of up to 16 associates and eight tutors
Regional Dean / SYSOPS (II) — Level II SYSOPS for your region	Regional short-circuiting to speed up intelligence flows as required
Multinational / SYSOPS (III) — Level III SYSOPS is IMC multinationally	

Note: SYSOPS = systems operations/ors

Figure 1.
IMC's electronic
networking with
two-level BBS

If addresses and telephone/fax data can be mistakenly construed as trivial, the quality and relevance of intellectual support provided to managers cannot. For many years faculty and associates worldwide have, in the best traditions of any growing multinational organization, claimed that all guidance was skewed towards the culture of those who were concerned, albeit on the entire network's behalf, to create it. Using the AIS we are all now able, at no greater interval than one month (and less if it is of major urgency), to comprehend, respond if necessary or simply observe the divergent situations making their own decisions on these matters. Those officers charged with assessing the quality of learning programmes within the network are indeed more able so to do by monitoring divergence and cultural adaptation than they were in the days when we all really pretended that everyone was doing the same thing.

And of course the great strength of understood diversity is that it triggers lateral thoughts in our own contexts. The Singaporean members of IMC evolved, along US lines for example, our pattern of adult Bachelor studies that was eventually taken up in Hong Kong, Macau and Finland, then South Africa and finally the UK. Each application was different but the confidence of success elsewhere encouraged change which, in each case, led the market concerned. What was true for the Bachelor programmes has been equally truse for doctoral studies and Masters both of Business Administration and of Philosophy in Training and Development. It has even been true for the reorientation of banking cultures from transaction focus to customer focus starting in England, spreading to Australia, then back to Ireland and thence to Malaysia (Wills, 1991). Each and every application was different but the network contribution was most substantial. In this latter instance it led to the launch by IMC's official publisher of what has since become the leading academic journal in the area, the *International Journal of Bank Marketing.*

Perhaps most fascinating of all has been the way in which the IMC network has responded to the challenge of "educational quality assurance". It certainly also spawned another very successful journal, *Educational Quality Assurance,* but it has brilliantly demonstrated the quality definition "fitness for purpose". In a traditional academic world, using peer-group review to arrive at a consensus of purpose, IMC's far-flung multinationalism shouted boldly that "purpose" had no end of definitions! MBA associates can, for example, be seeking to become more effective managers in the South African context of black economic empowerment; in a South Pacific island seeking offshore investors and World Bank finance; or in a first-world economy struggling for marginal market share improvements or at the leading edge of technology. As a network we are united in insisting that the questions are the same: how to be more effective managers? Yet our answers will be very different depending on the context in which we must be effective. The measure of quality will be how well they suit their contextual management purpose.

Financing the network

Throughout its history IMC's network has addressed its funding challenges with enthusiasm and some ingenuity! More than £30 million of sales have been generated, and the overall risks of the network have been broadly shared. Nonetheless, the overwhelming conclusion must be that networking still constitutes an almost insoluble major financing challenge.

If the network is not to be coercive each member must be self-financing and able to exercise independent judgement about participation. The challenge is most especially acute when any long-term investment issues arise. How can patient finance be mobilized without transforming the patterns of relationships within the network? How, in other words, can the benefits of equity finance be achieved in what is not perceived as a permanent enterprise? This constitutes a vital question for the future of all networks.

The answers that have emerged thus far (Kensinger and Martin, 1991) indicate that, while project financing is readily available and/or factoring of debtors, these do not assist long-term development. Major customers have sometimes helped here with long-term contracts to use as collateral for debt financing from banks, with settlements direct to the lending institution. Additionally, a number of instances of minority equity stakes have been reported, held by a few key clients. It is suggested that these can amount to sufficient funding in the effective network for all its needs – because networks expand or contract by definition to meet such changes as they confront. It does inescapably mean, however, that each major project or endeavour has to be financed de nouveau rather than by a board of management having shareholders' funds available to finance a portfolio of projects. It returns much of the top management function back to the marketplace.

Such an outcome can be either a major drag on development or a fillip depending on the competence and skill the network has at marshalling its case for project funding and delivering the targeted returns. Networking is, accordingly, an extremely harsh financial environment, and a high rate of failure must be envisaged. But networks are designed to be able to accommodate such problems, regrouping, reassembling, wiser than before about what can be achieved and how. Kensinger and Martin draw a close comparison between networking today and the medieval Hanseatic League.

IMC addressed these issues by establishing a not-for-profit central hub, with service contracts to three substantial network members who together had a wide spread of equity, i.e. we created a fluctuating equity-based partnership structure. These broad-ranging service companies took the risk out of the fluctuating fortunes of a wide spread of members right across the world. They were also able to act as a source of medium/long-term finance for the development of network assets. The official publisher, MCB University Press, which for a long while coerced the network, did so by providing the finance to create the action learning course materials and access to the body of knowledge on an up-to-date basis via the 130 journals which it publishes globally. Its offices worldwide also offered shared space, and it invested in a dedicated property in the UK. Five network members owned the publishing house between them.

The other two major sources of investment are geographically based – IMC (Europe and Africa) and IMC (Asia Pacific). They search for and encourage the development of locally based network members, earning a percentage share of sales for their efforts. Share ownership is open to any network member, and upwards of 40 participate currently. Most network members, however, are not highly capitalized and deploy what they do have available for the development of their own local network interests in a single country or even for a discrete product level within that country. Each finds the locus of investment with which he is most comfortable, and much debate takes place about the relative fairness of the margins and returns in the network.

At all levels there have been feast years and famine years, so no one level can claim that the allocations are unfair in any substantial way in the medium term. The balance of advantage, as in all channel structures, really lies in ensuring that the cake to be shared is first maximized. Much thought and a great deal of subsequent discipline has gone into seeking to ensure that this is the focus. The centre (SYSOPS as it is now called) is very lean indeed, and able to invest in the medium- and long-term development of the network only if sales are achieved among the grassroots members. SYSOPS income is volume related and must, accordingly, do all it can to facilitate volume sales, and repeat sales arising from quality services delivered.

On the half-dozen occasions when one or other network member collapsed leaving liabilities (now largely avoided through insistence on the bonding of programme fees), the best interests of the other network members one way or another have caused them to act to ensure that customers received full value as contracted.

Network roles and cultural evolution

Snow *et al.* (1992) characterize the managerial or leadership roles in networks as "brokering". Any individual's leadership has two clear components – the role structure and the effectiveness with which a given individual plays the allocated role. While the Atherton Intelligence System (AIS) developed in IMC certainly reinforces roles and facilitates communications between network members (see Figure 1), it can do no more than that.

Snow *et al.* perceive three broking activities. The first, and in many ways the most influential in the accomplishments of the network, is its architect. It is the true socio-entrepreneurial role described at the outset of this essay and, like all entrepreneurial processes, must be concerned for the success of absolutely every element within the network. The total architecture must, of itself, be a pleasure to behold, but the design must also be wholly functional for the purposes of those who live with it.

The architect coaxes others into the network – at all levels and in all manner of roles and relationship. They join wholeheartedly only if the vision is one with which they can identify and, if needs be, amend. Furthermore, because the environment in which the network grows and flourishes is necessarily so dynamic and so flexible, the role of architect is a continuous one – or certainly seems that way. Remember, the network has none of the accustomed financial stabilities of the equity-financed enterprise, with its patient investors. The network is only as successful as its last match! So architects must be restless, always and forever seeking to adapt to the emerging situation while at the same time giving a vital semblance of order for those seeking at any moment to lead in operational and caretaking roles in the network. These are the two other critical roles which Snow *et al.* identify. Their analysis of how all interact bears very close resemblance to contemporary analytical work on successful teams (Margerison and McCann, 1991).

The invisible architect for any network will always be its marketplace, its customers (Wills, 1993). But it is on the interpretation of what those phenomena require in product and service terms that Snow *et al.* are focusing. For IMC such architecture is encouraged out in the markets in discussion with customers. The seemingly apparent architects at dean and principal levels cannot really be said to hold that role. It is their role rather to unleash the network, not control it (Bush and Frohman, 1991). Any and every network member in contact with the marketplace and with customers must, within appropriate yet flexible and adaptive bounds, have the skill and the discretion to commit the network to a design. He must know that major new designs can and will be evaluated, evolved and supported by colleagues to a guaranteed service response level. He must know what constitutes a "substantive" and a "non-substantive" variation because, in the latter case, the network has the capability for virtually immediate response, with discretion at the customer interface. No network member sitting away from that interface may second guess what the customer needs or wants.

The challenge for the dean and principal is to make sense of the myriad market- and customer-derived communications inwards and to accommodate them as speedily as possible within the vision. Ironically the extent to which new ideas are readily accommodated within the vision depends on how clearly the vision is understood throughout the network at any node, at any moment in time.

The major misunderstandings and most heated debates arise between network members who have joined with extant institutional and cultural baggage that gets in the way of their embracing the network. It is the well-known phenomenon of culture clash which, for networks to survive, must be expeditiously and healthily resolved (Barnatt and Wong, 1992). Two network cultures cannot be sustained although that a new synthesized culture must often result when we are able to learn well from our arriving members. There are few more revealing critiques of any institutional framework available than those proffered by an incoming intelligentsia as they seek to understand why and how the structure they have joined actually works. They bring to their critique insights born of different cultures. The sentient network watches, listens and learns. So far as possible it accommodates as well, seeking the win-win synthesis.

This was well learned indeed in IMC when first the Canadian School of Management (CSM) then Business School Nederland (BSN) acceded to IMC's Common Multinational Academic Board. The CSM showed us how to provide action learning adult Bachelor programmes based on experiential learning already gained. The BSN proposed and spurred our development of the AIS worldwide and with it the concept of "action distance learning". Sets held fewer face-to-face meetings, overcoming the very real problems of regular attendances for busy managers, but we were nevertheless strongly sustaining the peer-group support process of action learning electronically.

The need to synthesize as you watch, listen and learn has been overwhelmingly driven home to us in South Africa. IMC's mandate there is to empower black faculty and black managers despite their lack of normally anticipated prior experience or education in all too many cases. The nation's history simply has not afforded a traditional preparation and its future is not prepared to, indeed could not, wait.

Such learning opportunities will forever appear at all network nodes, of course, provided that the network members have constructively joined the network rather than devoted their time to fighting a rearguard action. The greater the discretion accorded to the network nodes within a framework of trust, the more opportunities to learn will arise.

The formal functioning of the network is not, of course, the regular concern of its architects. The major functional roles are perceived as lead operator and caretaker. The first, the lead operator, certainly is a leader within the framework created by the architects, assembling the members to deliver and quality assure programmes. The caretakers are those individuals with especial concern to uphold and maintain the dignity and integrity of the network. They will be followers of the lead operators in the specific delivery of programmes but will have a unique role in developing and maintaining systems and sharing information, e.g. the AIS and IMC's Registry databases, facilitating the Ceremonial Congregations and Annual Professional Congresses, ensuring faculty training, induction, and five-year continuing professional renewal all take place. Ultimately the caretaker's emotions will be aroused most when disloyalty is perceived – when network members exploit the network for their own short-term gain at the expense of other actual or potential members (Schmidt, 1992). The caretaker here will point out the dysfunctionality present and seek disciplinary measures to protect its reputation. They are those who believe that a network can function effectively only on the basis of trust (Krackhardt and Hanson, 1993).

Successful lead operators are those who are visibly even-handed in their role play. Successful and acceptable caretakers do not overplay their hand. But IMC's 12-year history has been replete with examples of major successes and failures in both roles. Mature networks, and mature network members, can and do discuss and review how to overcome such dysfunctionality – on a regular basis – because the problem never goes away. It is endemic to networking no matter how many rules the mature network adopts.

Major causes of network failures

There can be little doubt that liberation management (Peters, 1992) has a powerful appeal in many countries across the world – and networking is what the liberated manager (or democrat) engages in. Nevertheless, there are major downsides to the construct. Miles and Snow (1992), who have been closely concerned as scholars with tracking the emergence and rise of networks, believe the most likely forecast is that their effectiveness will decline over time. Much of their reasoning has been trailed already in this essay, but the value of their contribution is that they codify the major causes of failure:

(1) First and foremost, any network has a life cycle for any given vision or set of visionaries. To continue it must change, either by evolution or with a discontinuity. Its maturing architects must know how to facilitate the quest for their successors.

(2) As networks become mature, the lead operator and caretaker roles, while vital to carry forward the projects concerned, will seek to institutionalize the network and constrain its market and customer responsiveness and flexibility. This, in its turn, will lead to disaggregation and loose coupling within the network unless the network is in a stable environment – which it seldom will be. Amidst calls for discipline the inspired networkers will frequently be heard either sounding the retreat or encouraging greater nodal autonomy to dissipate frustrations and release functional energies.

(3) Networkers who do not fully understand the nature of the processes involved will make two kinds of subtle mistake: first, the extension of networking beyond the limits of its capability and, second, the modification of its form in such a way as to modify its operating logic. IMC has many examples of these subtle mistakes.

(4) Dominant and coercive members are a necessary evil within any network. Successful networks only allow transitory coercion based on functional contribution to the network's purposes. They are quick to ensure heated debate when dysfunctionality occurs. Paradoxically, therefore, network architects need to be strong personalities but also extensive plagiarists of other member's ideas throughout the network. They must use situational role power to allow members close to the market and customers to flourish.

(5) Sub-networks, or "bow-ties", may develop within a network, engaging in secretive behaviours and excessive legalism, particularly where competition for markets and customers occurs. Successful networks must develop and uphold processes that avoid such dangers and respond speedily to whatever circumstances give rise to them in the first instance. Fairness, openness and trust, what Limerick (1992) has described as the *amicitia* of the network, must be upheld and be seen to be upheld.

Conclusions
Socio-entrepreneurial networks may well be an intangible asset so far as accounting convention goes, but their relevance as an organizational form for citizens of the twenty-first century seems inescapable. Not only do they

promise a liberated if tough workplace but they have the ability to regenerate themselves without anything like the trauma associated with the downsizing or collapse of the traditional equity-based corporation of the twentieth century. In an age of rapidly changing environments based on superior intelligence more rapidly diffused, this alone is enough to ensure a real future.

Nevertheless, there are many managerial challenges involved, many of which have been familiar in the literature and to practitioners for decades – networks simply aggregate them differently.

The human resource movement has championed the individual's self-development and contribution in the workplace for many decades (Limerick, 1992). Outsourcing, subcontracting, value-added channel management and procurement are all well-researched and understood issues. It is not in these areas, therefore, that the balance is most disturbed. It is in the fields of strategic purpose and funding that it is most considerably altered, where the discontinuities are greatest.

They are deeply interactive. The challenges of network financing are derived from the transitoriness of projects and tasks within the network vision. And strategic purpose is normally quite limited in its scope.

Sanity in the exciting chaos of networking will surely come to those architects who can best envision processes of rolling strategies and vision regeneration, with rolling funding to support them. The likelihood that a network can evolve into anything as stable as the traditional equity-based corporation is a contradiction in terms. If such stability is possible, the distinctive competences of networks will not command a premium. The architects must, without doubt, facilitate network funding if their visions are to be realized – either personally or in tandem with fellow networkers who clearly know how. The "hollow corporation", as the network has been called, needs the sap to rise just like all other human endeavour, or its leaves and branches will fall.

My own network, IMC, has chosen to confront its challenge of vision succession by rearranging its marketplace position. It resolved first, and as it transpired inappropriately, to link with a major managerial consultancy enterprise. This incident has been described already (Wills, 1992). Not deterred, it has now again sought out a major, potentially dominating partnership from the university sector to reflect the aspirations and expectations of the market and its customers. The wisdom gained from the previous failure has been deployed to seek a win-win relationship with the university sector. IMC's multinational network and its AIS electronic communications can greatly extend the work of any university; and IMC's work can at the same time be most considerably enhanced.

Only time can tell how well such strategic and financial envisaging will be. What is clear is that networks must use their ability to change shape to meet their market and customers' needs as the primary tool for securing their own futures.

References

Barnatt, C. and Wong, P. (1992), "Acquisition activity and organisational structure", *Journal of General Management,* Vol. 17 No. 3, pp. 1-15.

Blankenburg, D. and Johannson, J. (1992), "Managing network connections in international business", *Scandinavian International Business Review,* Vol. 1 No. 1, pp. 5-19.

Bush, J.B. and Frohman, A.L. (1991), "Communication in a network organisation", *Organisational Dynamics,* Vol. 20 No. 2, pp. 23-36.

Charan, R. (1991), "How networks reshape organisations – for results", *Harvard Business Review,* September-October, pp. 104-15.

Feneuille, S. (1990), "A network organisation to meet the challenges of complexity", *European Management Journal,* Vol. 8 No. 3, pp. 296-301.

Ghoshal, S. and Bartlett, C.A. (1990), "The multinational corporation as an interorganisational network", *Academy of Management Review,* Vol. 15 No. 4, pp. 603-25.

Huey, J. (1994), "The new post-heroic leadership", *Fortune,* 21 February, pp. 42-50.

Johannisson, B. (1987), "Beyond process and social structure: social exchange networks", *International Studies of Management and Organisation,* Vol. XVII No. I, pp. 3-23.

Kensinger, J.W. and Martin, J.D. (1991), "Financing network organisations", *Journal of Applied Corporate Finance,* pp. 66-76.

Krackhardt, D. and Hanson, J.R. (1993), "Informal networks: the company behind the chart", *Harvard Business Review,* July-August, pp. 104-11.

Limerick, D. (1992), "The shape of the new organisation", *Asia Pacific Journal of Human Resources,* Vol. 30 No. 1, pp. 38-52.

Margerison, C.J. and McCann, D. (1991), *How to Lead a Winning Team,* Team Management Systems, York.

Miles, R.E. and Snow, C.C. (1992), "Causes of failure in network organisations", *California Management Review,* Summer, pp. 53-72.

Peters, T. (1992), "Rethinking scale", *California Management Review,* Fall, pp. 7-29.

Pinchot, G. (1985), *Intrapreneuring,* Harper & Row, New York, NY.

Schmidt, D.P. (1992), "Integrating ethics into organisational networks", *Journal of Management Development,* Vol. 11 No. 4, pp. 34-43.

Snow, C.C., Miles, R.E. and Coleman, H.J. (1992), "Managing 21st century network organisations", *Organisational Dynamics,* Vol. 20 No. 3, pp. 5-20.

Theuerkauf, T. (1991), "Reshaping the global organisation", *McKinsey Quarterly,* No. 3, pp. 102-19.

Watson, M. (1990), "The networked organisation", *RSA Journal,* June, pp. 480-90.

Wills, G. (1991), "Enabling customers to drive your enterprise", *European Journal of Marketing,* Vol. 25 No. 4, pp. 199-216.

Wills, G. (1992), "Managing networking", *Scandinavian International Business Review,* Vol. 1 No. 3, pp. 52-70.

Wills, G. (1993), *The Enterprise School of Management,* MCB University Press, Bradford.

Chapter 10

Designing a quality action learning process for managers

The effective delivery of action learning requires the deconstruction of the normative curriculum to allow the customer to drive its re-creation from challenges that are meaningful and actionable in the learner's own context. This gives rise to concerns both among providers (who fear where customers might lead them) and among customers (who fear the responsibility they must assume for their own learning). And in the particular context of International Management Centres and the University of Surrey (Surrey IMC), where we are involved with learning at and from the workplace, there is a further concern among the intermediaries/brokers/funders of the process (who fear lest their backing of action learning is seen as a less worthy or credible process for the individual manager's career prospects than one driven by a normative curriculum).

Where such concerns are present, and they are not infrequent, one significant way to allay them is by means of visible, credible patterns of quality assurance. Providers, customers and intermediaries must all be reassured. This paper is a case analysis of International Management Centres' (IMC) journey of continuous improvement in this respect since 1982 when it metamorphosed itself from a professional association of Business School graduates (established in 1964 from what are today the Universities of Portsmouth, Westminster and Thames Valley) into a qualifying workplace action learning Business School awarding MBA, DBA and Practitioner Bachelor degrees wholly in its own right (Mumford, 1997).

The originating paradox
The 1982 metamorphosis was catalysed by research in the late 1970s specifically at Cranfield School of Management but more widely in industry and the management professions. It showed that, while student managers recognized the marketplace credibility of MBAs in their own career development, they criticized much of the tutorial as too theoretical. Employers for their part regarded it as deficient in skills to make use of the theoretical knowledge acquired. The paradox was that despite such disappointments MBAs were in great demand. Providers, finding themselves in such a seller's market, did little to respond to the concerns expressed.

IMC, because it was an association of graduates, felt sure something both could and should be done, and after an extensive search determined that the action learning approach pioneered by Reg Revans (1971; 1982), advanced in UK coal-mines and health care and then latterly in Belgium, was ideally suited.

With Molly Ainslie,
Journal of Workplace Learning,
Vol. 9 No. 3, 1997

A foundation consortium of industrial concerns and business school faculty initiated the first programmes in 1982. Support came from IDV/Grand Metropolitan and NatWest Bank in the UK, Dow Corning in Brussels, and from academics at Cranfield, Hull, Bradford and Queensland Universities. It was a dangerous enterprise academically if not for the customer or the broker. The deconstructed syllabus was not welcomed at a time when business schools had just set their chosen normative curriculum in concrete. However, success was assured by holding fast to the espoused customer focus and deploying those elements of the normative curriculum which were requisite for the re-created curriculum. Finally, at the conclusion of each degree programme the traditional pattern of external examination was conducted. Quality in these early days (Peters, 1988) was seemingly assured by our dedication to customers' needs and by the determination within each programme (known as a Set) by all concerned to expect "quality" on a fitness for purpose criterion. Great effort was made to find tutors with the facilitation skills necessary in adult learning and to empower the Set first to counsel, then if necessary to remove, poor performers. Great attention was paid to the design and conduct of all assessed assignments to ensure that they reinforced the goals of actionable learning – that they benefited the career of the individual manager concerned and that they gave a measurable return on investment (ROI) for the sponsoring organization (Ball, 1991; Wills and Oliver, 1996).

Nothing succeeded like success. A glittering array of corporate names joined with the originating enterprises – Cummins Engines, Jones Lang Wootton, Shell, ICI, Midland Bank, Malaysia Airlines, Ernst & Young, and more. And they were seeking our programmes across the world – in Australia, South Africa, Hong Kong, Singapore, Malaysia, Indonesia, Papua New Guinea, Finland, Holland, Vanuatu and elsewhere. Our lines of communication became rapidly extended and processes of quality assurance became a very high priority indeed (Kozubska and Wills, 1992). Our awareness of this need had been well highlighted in our successful submissions for accreditation by the British Accreditation Council and by the Washington-based Distance Education and Training Commission in 1985 and 1986.

Action learning's Achilles' heel
The marketplace success of a customer-focused workplace curriculum did not surprise us. We had anticipated it but our potential competitors, being in a seller's market, had neither the incentive nor the will to follow us until more than a decade had passed – rather they characterized us as oxymoronic and accordingly unworthy.

Yet the founders of IMC's action learning initiative were all products of the normative curriculum and were well aware of its particular benefits that action learning clearly did not offer. By the latter's determined focus on actionable customer-focused issues for the re-creation of the curriculum, a helicopter view of the totality of the body of learned knowledge was not imparted and could not be obtained. Issue-driven learning left gaps, often quite substantial, that no thoroughness in attention to scholarship in the issue areas could avoid. And this was even after the fullest involvement with fellow action learners in Set discussions of one another's actionable issues, which of itself extended coverage greatly beyond the individual's own focus areas (Sutton, 1990).

We accordingly resolved from the outset to address this in two ways: the first, which has already been alluded to, was the inclusion in the programmes of "courseware" covering the all-significant areas of the normative curriculum, albeit for requisite use with total exposure only by browsing. Second, however, we resolved to place great emphasis on ensuring that all action learners were aware of and became advocates of "understanding how we learn" and how to improve our learning styles (Mumford and Honey, 1986). This determination was institutionalized in myriad ways – through faculty training and induction, learning styles testing and action planning, evaluative assessments of learning, both at workshops to talk over learning log entries and as credit-earning outputs (Thomas, 1993). This attention was further continued after graduation through one- and five-year continuing renewal assignments leading to Companion Membership (Mumford,1996).

In this manner we ensured a platform of learning and independent action for graduate managers to build on, as issues arose outwith their specific issue-driven action learning.

Testing our faith

The 1988 UK Educational Reform Act was a supreme test of our faith in the action learning process and our customers' right to join with us in re-creating the curriculum. The academic establishment, encouraged by Government, which deliberately reinforced the cartelization of higher education in the Act to eliminate magnified malpractices, sought to return our curriculum, via validation processes and legal sanctions, to its normative view. But we had seen sufficient of action learning's "fitness for purpose quality" and the enthusiastic acceptance of our educational approach by customers to be well willing to suffer the consequent slings and arrows of such outrageous fortune – these included a commercial débâcle and subsequent refinancing of our espoused goals by faculty and graduates in 1992 from their own pockets.

IMC was especially sustained by its corporate clients but most particularly by adherents in the Asia Pacific, Holland and South Africa, where traditionalist cartels were then less firmly entrenched. And so from the outrageous fortune a re-engineered global institution was sensibly forged. This provision of action learning programmes in myriad countries then gave us an excellent basis for comparative benchmarking of quality achieved. Best Graduates at Bachelor/ Master/Doctor, Best Academic Partner, Best Companion Member, Best Learning Organization awards were competed for globally and honours shared around the world. And twice each year, at graduation workshops and Annual Professional Congress as many as might met together at a single location – Helsinki, Amsterdam, Johannesburg, Curaçao, Hong Kong, Brisbane, Kuala Lumpur, London – to share and compare. At one of these, Australian and Dutch colleagues were most determined that we should come to terms with the advancing electronic communications systems for straightforward issues like the Multinational Registry and for interactive communications. The initial proposals were for electronic data interchange (EDI) and bulletin board systems (BBS), both of which went live in 1994.

Evaluative Research Fellowships were also set in place each year, one of which (Peters, 1995) quite specifically explored whether IMC could usefully employ ISO 9000 approaches in its customer-focused action learning vision. The overriding message from Peters, who heads IMC's School of Quality Management, which offers an Executive Diploma/MBA in Quality Management in association with the British Standards Institution (BSI), was readily anticipated. The challenge was how to engender the enormous energy required to develop and implement an ISO 9002 approach globally. As with all commencing such an endeavour, many of the component elements are present but they do not have the coherence that a total quality management orientation affords (Peters and Wills, 1996) (see Appendices 1 and 2).

Accessing requisite knowledge

We were determined to be true to our action learning, customer-focused vision as we advanced. A normative construction could not be acceptable and many doubting Thomases proffered the view that ISO 9000 processes often collapsed under the weight of their own bureaucracy. We resolved to evaluate the voluminous literature on ISO 9000 and in particular the experience of educationists – a process we have since continued on a dynamic basis. Here we shall focus in particular on the latter, which is relatively scant.

Idrus (1996) offers a comprehensive global review of TQM in the educational sector. While New Zealand, where he is Director of Otago Polytechnic, disappoints him, he found considerable evidence of successful applications in the USA and Europe but was critical of business schools that advocate TQM without taking their own medicine. That hurt. His overall conclusion was that, unlike the inner drive within those coming from action learning, the majority of cases of implementation that he identified arose from outside pressures. This is exemplified in Asser and Haines (1995) at Oxfordshire County Council; Cave *et al.* (1995) at Brunel University; Doherty (1995) at the University of Wolverhampton; Hill (1995) at Queen's University, Belfast; Rippin et al. (1994) at Bolton Institute; and Sommerville (1996) at Swanley College. In almost all these circumstances it was the pressure of external competition either for Government resources or for customers that drove the initiative forward. Most contributions allude to the Governmental desire to measure the performance of education whether via performance efficiency indicators or via input/output measures. The influence in the UK from the now defunct Council for National Academic Awards (CNAA) was profound. Throughout their 25-year history as the accrediting body for new higher education institutions they had built up and implemented a disciplined regime towards what they perceived as quality (CNAA, 1992a). In their last years they also began to embrace the customer-derived contribution to the quality debate (CNAA, 1992b) which has since gathered force globally.

Marsh (1991) captured the dilemma as elegantly as any author when he observed: "educational institutions should neither embrace the discourse and practices of the commercial quality culture wholeheartedly nor should they embrace them uncritically". The creation of a total quality management environment that monitors nonconformities and calls for action demands the involvement of individuals in all internal processes that have the clear purpose of improving the academic quality of teaching, learning and the student experience. The challenge is how to bring this about in the typical higher educational structures without relying (exclusively) on external pressures. Vision from the top is readily offered as an effective way to achieve most organizational and cultural changes and the educational sector proves to be no exception (Schoengrund, 1996; Sommerville, 1996).

It would be unfair, however, to suggest that a considerable volume of activity and reported research was not taking place on a programme-by-programme basis – oblivious to external pressures or top-down vision. There seem to be three strands. The first is either ISO 9002-driven (McRobert, 1995) or builds on the traditions of quality circles and teams (Asser and Haines, 1995; Boaden and Dale, 1993; Collins et al., 1991; Goulden, 1996; Green, 1996; Morgan, 1990). The second, expressed by Morgan and Piercy, (1992) and O'Neal and LaFief (1992) urges marketing to enforce the customer-driven focus of the quality movement. But this strand goes further in studies that take a consumerist view, looking at the learner's own perspectives (Muller and Tonnell, 1993), focusing on the individual student as the true and direct driver for quality (Showalter and Mulholland, 1992) and even more interestingly via a simulation exercise (Bacon et al., 1996). This latter example is a powerful way for helping students to learn how to learn and is worthy of considerable development. Schoengrund (1996) also explored the multilevel customer/consumer patterns in education which have strong relevance for workplace learning, as already indicated.

The third strand has been labelled quality function deployment (QFD). Pitman et al. (1996) applied it to MBA programme design in North America and assert that it "helps ensure the voice of the customer is clearly heard and followed in the development of a product or service". Originating in Japan, it works with cross-functional development teams to explore the Whats and the Hows and uses graphical displays to guide the process. Learning processes and curriculum compatibility are readily susceptible to such approaches (Lo and Sculli, 1996). Alitalia (Ghobadian and Terry, 1995) have made extensive use of it for customer service educational programmes, as have real estate agents (Campbell, 1996).

Finally, our review of the published body of knowledge highlighted a number of other seemingly discrete fields of research and evaluation that contributed to our own thinking. Morgan (1990) looked at management activity issues that assisted the quality focus for our Registry system. Hosie (1995) describes quality assurance processes in his analysis of human resource information systems that have greatly figured during the development of our faculty and Graduate databases for online searching against meaningful criteria. This database approach in particular highlights the need for faculty development and their progress along that journey with appraisal feedback and mentoring and ultimately giving rise to our own continuing renewal process as Companion Member, which is similar to Farrugia's model (1996).

Himmett and Knight (1996) most gratifyingly explored the determinism of the assignments set for any learning experience and agreed strongly that they must reflect the learners' needs not those of the normative hierarchy. Cleary (1996) reinforces the point again, linking it with how students learn, using recognizable learning cycles but also working from disparate learning styles.

Focusing on ISO 9002

The dynamics of our learning approach, the exhortations of the Head of our Quality Management School, and the globally located delivery sites for our programmes readily came together to justify and give us the resolution to proceed with ISO 9002. We recruited one of our BSI collaborators as our consultant, and Registry and Bursar colleagues, led by Molly Ainslie, came to terms with what was going to be required. Like so many before us, we quickly came to understand that we had to delineate and communicate the managerial processes of our business school that much more effectively, so that all involved with us – upwards of 300 Faculty, 4,000 Graduates and 400 current students (known as associates) – could discern what they were entitled to expect. And then we had to measure our nonconformances and take deliberate actions either to emend the elements of quality or to deliver them more effectively. It took nine months for the quality team to complete the Internet site and associated documentation to win approval for the process in action – at their first attempt. Charters were crafted for associates, faculty, graduates and partners that, by making use of Internet hyperlinks to our extensive site, were a major advance on the old paper systems and BBS (Ainslie, 1996).

Perhaps most significantly, the search engines created and their criteria meant that information that only a few previously knew to exist could be speedily traced – and remember we are a global business school which had always exacerbated the problem. Almost as important, however, was our ability to keep all the information up to date. This included course materials which were formerly, on average, two and a half years old now being never more than six months since last revision.

The distribution of paper work from the Multinational Registry and Common Multinational Academic Board was replaced by Web Forum/e-mail proactive notification and user downloading. This ranged from full programme resources at one extreme to the Minutes of Statutory Meetings, Annual Congress outcomes and research findings. Paper was not eliminated but printing was done on demand at the point of use, eliminating global postal expense and delays as well as increasing awareness. The sophisticated Web Forum/Virtual Academy framework now evolved allows not only for discourse among programme participants but globally for the Senior Tutor, Chairmen of MBA, Doctoral and Bachelor programmes, for the Annual Congress and all major meetings – these latter run as "virtual" discussions of agenda items for six/nine weeks prior to the face-to-face event and enable those unable to attend physically because of geographical distance to participate actively nonetheless.

A fillip for the process

The analysis of nonconformances for ISO 9002 readily showed that much of the infrastructure required for effective management depended on others around the globe to supply data and to populate the sites concerned. We had previously re-engineered our Registry, in pre-ISO 9002 days, to use electronic data interchange (EDI) from distributed sub-Registries in Australia, Malaysia, South Africa and Holland to great effect as opposed to awaiting their transmission of paper-based records of application data, grades and fee settlements. We resolved to stay with the EDI strategy after ISO 9002 accreditation to create a wholly dedicated resource at the multinational monitoring point which would proact strongly with the sub-Registries and vice versa.

Furthermore, the Internet resourcing of our programmes involved the creation and ongoing maintenance of virtual professions cum academies (Teare, 1996) in partnership with client enterprises, e.g. BAA plc and Fina, as well as regionally, e.g. Asia Pacific Management Forum with its sub-Forums for Japan and Australasia. The creation process eventually evolved a series of modular templates, many of which could be routinized, but the challenge of ongoing maintenance and dynamic activities is and will for ever be much more problematic. It calls not just for monitoring of course but also for self-starting motivation. Here, as with the learning process itself among associates, the emphasis had to be placed on induction, training and group processes to achieve the requisite inspiration and desired fitness for purpose.

From fillip to imperative

IMC's single-minded dedication to action learning in the workplace since 1982, together with its leadership position as an Internet-driven institution, represents considerable strengths in the contemporary, turbulent educational world. The external pressures, well rehearsed in the literature reviewed and referenced, are encouraging universities at large to look ever more closely at the opportunities available in either or both of these areas – and globally, where once again we have unique strengths some 15 years up the learning curve with many challenges addressed and overcome to show for it.

Accordingly the joint venture which emerged in September 1996 between the University of Surrey and IMC was not, on reflection, surprising. There were very considerable mutual advantages to be gained by both parties if they could agree on how to build on what had clearly been learned to be effective in global workplace learning, and what was being learned on the Internet – without damaging the vital reputation for "quality" enjoyed by the university within the Government-funded sector of higher education.

IMC's non-traditional patterns of evaluative research and its strong emphasis on learning to learn constituted a good starting-point. But problems were obviously present in terms of concern for sponsored in-post managers and traditional entry criteria. Some of the most innovative designs for qualification awards, e.g. Practitioner Bachelors and Doctorates, Master's awards for Faculty Development and Post Qualifying evaluative awards called A+ and Companion did not harmonize with the existing university framework. Validation panels for concurrent award schemes were unable to comprehend and evaluate the leadership role IMC had gained on the Internet. University faculty, schooled in scholarship as an appropriate end-in-itself and an *ad hominem* approach to teaching, were not always accepting of the workplace realities of focused learning and team reliance.

While IMC, which includes many current and former university academics in its faculty, would not presume to suggest how the university should operationalize its scholarly enquiry, the purpose of the joint venture was for IMC to show the way for the university to be yet more actively engaged in the workplace and with Internet resourcing. As can well be imagined, the discourse on quality in terms of fitness for purpose speedily focused on "whose purpose?" and, as that matter clarified, albeit gradually, the requirements for quality action learning from the joint venture took on their own shape.

There was ready agreement that the EDI modelling for global sub-Registries should be adopted from IMC. It was also concluded that, once assurance was gained for the core curriculum resourcing of programmes (the courseware, as it is known), the design and implementation strategies offsite with corporate or regional theming were areas where IMC's learning was well advanced. And then the same resolution was also quickly reached in relation to faculty skills as action learning facilitators.

Indeed, the reliance on IMC to show the way, which in retrospect should not have been particularly surprising, began to have the potential to make the joint venture one-sided. The *raison d'être* for the university entering into the joint venture was not to have IMC as a stand-alone unit but to forge links, share understanding across the university, and enable many others to evolve and develop in these areas. Failure to achieve this could be expected to undermine the joint venture's support, which from the outset was at the highest level, indeed unanimous at Senate. Accordingly, Surrey IMC resolved to:

(1) make dynamic quality assurance of the action learning processes our prime focus; and

(2) involve as many other university faculty as possible in the processes beyond the immediate identification with management learning.

Dynamic quality assurance
The quality assurance focus taken is conducted on the Internet. It is accessible at all times at the university or throughout IMC globally by authorized individuals. It encompasses what are deemed to be the major elements of a

quality action learning process and identifies the specific Registry elements involved.

The dynamic or continuous improvement of the system lies in the requirement that via EDI/online editing rights, the sub-Registries globally must update all elements monthly. Nonconformances are monitored and corrective action is taken.

While none will underestimate the challenge of keeping the information up to date, without the ISO 9002 disciplines and framework we would have been quite unable to respond as we have. But its elegant, visible success paradoxically created much hubbub among faculty members. After many years of constructive and less than constructive criticism of the management of these procedures, the arrival of a state-of-the-art system caused concern too. Surely too much attention had been given to the administration at the expense of the true seat of learning, namely within the Set with the associates.

This was, of course, a happy message for Surrey IMC's Senior Tutor who, with the Deans globally in support, is charged to ensure that all faculty (IMC and Surrey) who wish to tutor on joint-venture programmes are professionally competent and refreshed in the processes used. The opportunity was rapidly taken to demand a higher level than hitherto of professional skills from all faculty.

For all who were acknowledged action learning practitioners two things were henceforth required:

(1) immediate refreshment and Internet induction; and

(2) unless already qualified as an educationist, enrolment on Surrey IMC's own Postgraduate Advanced Diploma/MPhil programme with the benefit of a Faculty Development Scholarship forthwith.

For those who were not yet action learning practitioners, and this included many from the University of Surrey, the refreshment phase at (1) above was replaced with a two-day induction/ novice tutor period, with (2) still required.

The further key dimension to faculty professionalism that is being addressed is the leadership and design of total programmes with corporate clients. Here a Delivery Effectiveness Set has been established with faculty from the university and IMC to seek to share, transfer and improve all-round skills in areas as elusive as "client management" and programme budgeting. This theme has also been made the focus for the global 1997 Annual Professional Congress to be held in The Netherlands. It honours what evaluative research has shown again and again, in particular that the supportiveness of the workplace environment, when and where action learning is in progress, is one of the most significant determinants of success. Steering committees have a vital part to play, along with suitable mentoring procedures, to achieve this. Mentoring training is always offered as an integral element in all programmes provided.

Finally, and this is a matter of most considerable importance, IMC Faculty are seeking to build closer relationships with the University of Surrey Faculty in their discrete subject areas so as to be at the forefront of academic thinking/ expertise as well as in the deployment of knowledge in the workplace. Such discourse can, it is hoped, overcome the inherent weakness of action learning, which is that it frequently has no ready access to the extant body of knowledge; which is not to gainsay the new developments in literature searching online, but rather to emphasize the enduring value of debate face-to-face. The university's strength as a centre of academic excellence is greatly welcomed.

Lateral participation

The second element of strategy in Surrey IMC's search for quality assurance within the joint venture is lateral – outside the immediate field of management. Here there has been exceptional good fortune from the outset in that the university is in the process of quite deliberately re-engineering its adult, continuing education programmes. These have swiftly become the focus for Internet Virtual Forums and for action learning focus development – ranging from art history to religious studies.

Additionally, the active knowledge archive services of IMC's official publisher, allied with its global authors' club known as Literati Club, has evolved a Surrey IMC Authors' chapter similar to that already in place for IMC (Day, 1996). All University doctoral students and research staff have been invited to join the Surrey IMC Authors Internet grouping which matches them up with at least one appropriate journal editorial /review team and gives continuous pre-print and peer refereeing feedback as articles progress towards final stages for publication. At a time when universities are expected not only to deliver quality tutorial contributions through well designed learning frameworks but also to advance, and be seen to advance, scholarship, the Surrey IMC Authors service is of exceptional benefit to both parties.

In the spirit of continuous improvement: research implications

Surrey IMC, by bringing the two polarities in the educational quality assurance debate into one focused context – workplace learning – offers the potential for considerable continuous improvement and a forum for evaluative research. Not that one can or should triumph over the other. Universities must always remain the guardians and catalysts for scholarship. But if as institutions they also wish to espouse market/customer-driven learning, and lifelong learning, they must also come to terms with and respect the workplace context and leverage it via appropriate media, including the Internet, in the cause of more effective higher education.

To be criticized as oxymoronic is under such assumptions to miss the point. The higher education challenge is to do both well, attaining the highest achievable quality levels of fitness for the twin purposes. Research must be clear as to which polarity is appropriate and outcomes measured accordingly.

IMC's monotechnic approach for the past 15 years has enabled it to hone many powerful workplace learning approaches but it has lacked the especial strength that can be derived from a joint venture with a scholarly university institution. Equally, ambivalence about the legitimacy of workplace learning and its perceived threat to scholarly endeavour have meant that universities have often failed to serve a very large client group that greatly admires but frequently despairs of them.

To quote Marsh (1991) again: "these differences of perspective can easily become caricatures and the resultant stand-off between the two partners ultimately does nobody any good". The Surrey IMC joint venture is no ideological stand-off. It hurts and it creates heat and lively debate and is consequently genuinely important as we try to keep what works and discard what does not. And, after two years working thus together, we are committed to standing back to identify a worthwhile five-year horizon. Fitness for respective purposes can be expected to prevail.

References

Ainslie, M. (1996) "ISO 9002 quality charters", to be found @ http://www.imc.org.uk/imc/news/iso9002/quality.htm

Asser, M. and Haines, J. (1995), "A quest for the best", *International Journal of Public Sector Management,* Vol. 8 No. 7, pp. 6-14.

Bacon, D.R., Stewart, K.A. and Giclas, H. (1996), *Journal of Management Education*, Vol. 20 No. 2, pp. 265-75.

Ball, C. (1991), "Learning pays", *Education + Training*, Vol. 33 No. 4, pp. 4-5.

Boaden, R.J. and Dale, B.G. (1993). "Teamwork in services: quality circles by another name?", *International Journal of Service Industry Management,* Vol. 4 No. 1, pp. 5-24.

Campbell, J. (1996), "From bricks and mortar to service excellence", *Business Quarterly,* Summer, pp. 65-9.

Cave, M., Harvey, S. and Henkel, M. (1995), "Performance measurement in higher education – revisited", *Public Money and Management,* October/December, pp. 17-23.

Cleary, B.A. (1996), "Relearning the learning process", *Quality Progress,* April, pp. 79-85.

CNAA (1992a), *Evaluating the Quality of the Student Experience*, CNAA, London, February.

CNAA (1992b), *Academic Quality in Higher Education: A Guide to Good Practice in Framing Regulations,* CNAA, London, July.

Collins, D., Cockburn, M. and MacRobert, I. (1991), "Sandwell College: provider of quality assured education", *Quality Forum,* Vol. 17 No. 3, September, pp. 126-8.

Day, A. (1996), *How to Get Research Published in Journals,* Gower Press, Aldershot.

Doherty, G.D. (1995), "Accountability and excellence in education", *Total Quality Review,* January/February, pp. 37-44.

Farrugia, C. (1996), "A continuing professional development model for quality assurance in higher education", *Quality Assurance in Education,* Vol. 4 No. 2, pp. 28-34.

Ghobadian, A. and Terry, A.J. (1995), "How Alitalia improves service quality through quality function deployment", *Managing Service Quality,* Vol. 5 No. 5, pp. 25-30

Goulden, C. (1996), "Supervisory management and quality circle performance: an empirical study", *Benchmarking for Quality Management & Technology,* Vol. 2 No. 4, pp. 61-74.

Green, D. (1996), "A case for Koalaty Kid", *Quality Progress,* August, pp. 97-9.

Hill, F.M. (1995), "Managing service quality in higher education: the role of the student as primary consumer", *Quality Assurance in Education,* Vol. 3 No. 3, pp. 10-21.

Himmett, K. and Knight, P. (1996), "Quality and assessment", *Quality Assurance in Education,* Vol. 4 No. 3, pp. 3-10.

Hosie, P. (1995), "Promoting quality in higher education using human resource information systems", *Quality Assurance in Education,* Vol. 3 No. 1, pp. 30-5.

Idrus, N. (1996), "Towards total quality management in academia", *Quality Assurance in Education,* Vol. 4 No. 3, pp. 34-40.

Kozubska, J. and Wills, G. (1992), "Total quality assurance on IMC programmes", 4th IMC Annual Professional Congress, @http://www.imc.org.uk/imv/news/daproces/dap20.htm

Lo, V.H.Y. and Sculli, D. (1996), "An application of TQM concepts in education", *Training for Quality,* Vol. 4 No. 3, pp. 16-22.

McRobert, I. (1995), "Hermeneutics and human relations", *Total Quality Review,* January/February, pp. 45-52.

Marsh, P. (1991), "Bounce in the showroom", *Times Higher Education Supplement,* 1 November.

Morgan, M. (1990), "Quality circles: management accounting applications", *Management Accounting,* November, pp. 48-51.

Morgan, N.A. and Piercy, N.F. (1992) "Market-led quality", *Industrial Marketing Management,* Vol. 21, pp. 111-18.

Muller, D. and Tonnell, V. (1993), "Learner perceptions of quality and the learner career", *Quality Assurance in Education,* Vol. 1 No. 1, pp. 29-33.

Mumford, A. (1996), "5-year continuing review: progress to 1996", *Jubilee Fellowship Report,* @ http://www.imc.org.uk/imc/news/occpaper/9apr96.htm

Mumford, A. (Ed.) (1997), *Action Learning at Work,* Gower Press, Aldershot, being an anthology of 30+ evaluative studies of IMC action learning activities since 1982.

Mumford, A. and Honey. P. (1986), *Using Your Learning Styles,* Honey, Maidenhead.

O'Neal, C.R. and LaFief, W.C. (1992), "Marketing's lead role in total quality", *Industrial Marketing Management,* Vol. 21, pp. 133-43.

Peters, J. (1988), "Customers first: the independent answer", *Business Education,* Vol. 9 No. 3/4, pp. 34-41.

Peters, J. (1995), "Quality assessing the virtual business school", *David Sutton Fellowship Report,* @ http://www.imc.org.uk/imc/news/occpaper/sutton.htm

Peters, J. and Wills, G. (1996), "ISO 9000 as a global educational accreditation structure", State Department Conference, Washington DC., for The Center for Quality Assurance in International Education,9 May @http://www.imc.org.uk/imc/news/occpaper/washingt.htm

Pitman, G., Motwami, J., Ashok, K. and Chun, H.C. (1996), "QFD application in an educational setting", *International Journal of Quality & Reliability Management,* Vol. 13 No. 4, pp. 99-108.

Revans, R. (1971), *Developing Effective Managers,* Praeger, New York, NY.

Revans, R. (1982), *The Origins and Growth of Action Learning,* Chartwell Bratt, Bromley.

Rippin, A., White, J. and Marsh, P. (1994), "Quality assessment to quality enhancement", *Quality Assurance in Education,* Vol. 2 No. 1, pp. 13-20.

Schoengrund, C. (1996), "Aristotle and total quality management", *Total Quality Management,* Vol. 7 No. 1, pp. 79-91.

Showalter, M.J. and Mulholland, J.A. (1992), "Continuous improvement strategies for service organisations", *Business Horizons,* July/August, pp. 82-7.

Sommerville, A.K. (1996), "Changing culture", *Quality Assurance in Education,* Vol. 4 No. 1, pp. 32-6.

Sutton, D. (1990), "Action learning in search of P", *Industrial and Commercial Training,* Vol. 22 No. 1, @ http://www.imc.org.uk/imc/news/daproces/dap16.htm

Teare, R. (1996), "The dynamic curriculum: a prospectus for organizational learning", *IMC Joint Atherton & Sutton Fellowship Report,* @ http://www.imc.org.uk/imc/news/bulletins/fellow.htm

Thomas, J. (1993), "Researching learning to learn", @ http://www.imc.org.uk/imc/news/daproces/dap13.htm

Wills, G. and Oliver, C. (1996), "ROI, Measuring the ROI from management action learning", *Management Development Review,* Vol. 9 No. 1 pp. 17-21,@ http://www.imc.org.uk/imc/surrey. uni/papers/roi.htm (see Chapter 8).

Appendix 1. The process of continuous quality assurance

The Joint Academic Board of Studies accomplished this as follows:

1. *Aptitudes and skills* together with prior learning and experience and corporate references will be evaluated and assessed prior to entry by any Associate and this information (a) collated in the registry database (b) used as a proactive guide to action learning tutorial support and discussion in the Set and (c) to facilitate company-specific support.

2. Throughout their programme the client corporation and the Joint Academic Board of Studies maintain an advisory and evaluative relationship via a formally constituted Steering Committee, chaired by the Director of Studies for the particular pattern of action learning studies agreed. It meets before commencement and thereafter three monthly.

3. Throughout their action learning studies all Associates maintain and review a *Learning Log* which is subject to midterm evaluation in the Set and to a final assessed assignment known as the *EAML*.

4. *ISO 9002 Associates' Quality Charter* rights and responsibilities are monitored and any nonconformance addressed.

5. Patterns of *internal examination* and evaluative feedback to *agreed service levels* throughout the action learning process provide a strong basis for Associates' learning and improvement.

6. The *appraisal* of all tutorial inputs and the *counselling* of individual Associates by the *Set Adviser* provide opportunities for adjusting to meet nonconformance with the standards required from both parties to the learning contract.

7. The *external examination process* ensures comparison of outcomes with other routes to awards and objective feedback continuous *improvement of design and delivery*.

8. The quality assurance intelligence gathered at 1-7 above is tabled quarterly at the *QA Home Page* thereby giving transparency to the process throughout the 24 months.

9. The *evaluation* of the *Virtual Academy Forums* is achieved proactively via ListServes to the Validation & QA Group.

Appendix 2. Dynamic Quality Assurance Report

This Home Page gives access to the Continuous Quality Assurance process for all patterns of action learning studies followed in the University of Surrey IMC with quarterly updating by the Multinational Registry Executive under ISO 9002 procedures.

1. *Sets in progress via their Home Pages*

    1.1 *Associates and their qualifications*

    1.2 *Job descriptions*

    1.3 *Schedule for assignments and grades achieved*

    1.4 *Set meetings*

    1.5 Set Adviser and tutorial faculty CVs

    1.6 Learning Log monitor and EAML outcomes

    1.7 Quarterly Set Adviser Report

    1.8 *Quarterly Steering Group Report*

    1.9 Approved External Examiner

1.10 *Specific Brochure*
2. *Associates' Quality Charter*
    2.1 Questions raised and action taken
    2.2 Variance reports and actions taken
3. *Faculty Appraisals and Actions Taken*
    3.1 For each course tutor
    3.2 For Set Adviser
4. *Graduation Workshop Reports*
5. *External Examiners' Reports*
    5.1 The Reports
    5.2 Action taken as required
6. *Virtual Academy Forums*
    6.1 Creation updates and content development
    6.2 ListServes Usages and feedback
    6.3 External Assessor Comments

Chapter 11

Creating a marketing intelligentsia

More intelligent marketing activities

On the far side of technology-driven database marketing thinking, there is a largely unexplored territory for more intelligent marketing professionals. It is the up-side of George Orwell's down-side 1984. It certainly is territory where a goodly proportion of the waste, which we all know inevitably goes into traditional above the line promotion activity, can be eliminated at a cost which increasingly looks beneficial. A level of personal precision, of personal segmentation, for each customer that can replace impersonal surrogates, has become attainable. But the issues which the emergence of databases raises for marketing go far beyond promotional cost benefits.

Accordingly, it is time the electronic database phenomenon entered the total marketing conceptualization and thence the marketing textbooks. The dilemma is where to put it.

Kestnbaum[1] set the tone a decade ago when he asserted:

> We are capable of rendering a professional personalized service because we gather marketing information into a database. We have to show customers clearly and dramatically the value of our having that information for them as well as ourselves. It is potentially no different for the personal salesman in our favourite store who has known us for 20 years, who knows our tastes and the products we like.

The salesman in our favourite store has intelligence about us as customers and as individuals and in the aggregate. He knows what items will be repeat purchased, and what switching between brands may occur. He has a shrewd idea of what new products are likely to be attractive to each one of us and what additional services may be provided. He knows our price preferences and he can always ask us for our opinions on recent, current or future circumstances. He knows who among us shop early and who shop at the last moment, and what interpersonal communications approach we prefer. He does not intrude on our privacy as he relates with us but nonetheless gets his message across.

Since the advent of commercial radio and television advertising, and the emergence of mass merchandizing in megastores, we have at times forgotten him. His salary costs have frequently rendered him less cost-effective to his employers than an impersonal approach. But now the wheel is turning and much of what he offered can be accomplished electronically. That must go into our textbooks.

With Bev Bruce and Timmie Duncan,
Marketing Intelligence & Planning,
Vol. 9 No. 4, 1991

There are facets such as face-to-face relationships that electronics cannot yet match. But, when it comes to memory, provided we have captured and updated our data accurately, the database will be well ahead. And as for aggregation/ disaggregation routines at speed, we are in a different era altogether.

Figuring that our favourite salesman could never accomplish in time to be beneficial is accurately available on a real-time cumulative basis.

Such new-found ability to know, understand and respond to customers with hitherto unavailable alacrity is the second major issue for the textbook. It takes the concept of timeliness of information a quantum leap forward. Nowhere is this more readily discernible than in sales feedback on new product launches and direct promotional campaigns.

What is good for our existing customers is certainly also good for customer prospecting, a key area of marketing investment. Database matching guides marketing activities to expand the profitable customer basis. It pinpoints the potentials, qualifies them to avoid as much wasted effort as possible, and provides the integrated systems to plan, progress, monitor and control the prospecting process. This too must find its way into our textbooks.

The same disciplines that have been described above for our own dedicated salesforce with existing or prospective customers can be readily extended to initiate and build our channels of distribution. The understanding it permits the prime supplier to have of the total channel facilitates greater operational discretion for the distributor. But it also goes much further. Once the distributor perceives the benefits for himself and creates a customer actual or prospect database he can also move forward to create an integrated supplier database that permits comparative shopping against product specification, availability, after-sales support, price, settlement terms and the like.

Marketing research assumptions too are transformed. The ability to structure a representative sample of customers and prospects from the marketing database is readily available. And the level of sophistication in terms of segmentation that can be brought into play outstrips almost anything that can be contemplated in a random or quota field survey by a research organization not in possession of such a database. The implications of this aspect for the future structure and development of an independent marketing research industry are profound indeed. It impacts on every research area from attitude measurement and retail auditing to product concept testing, pricing research and test or pilot marketing.

Perhaps most interesting of all, however, it embraces the marketing research gathering processes into a total framework that monitors subsequent behaviour far better than cumbersome panel methodologies have traditionally aspired to do.

A new marketing research industry has emerged, and grows apace, of data brokers, lenders, laundry and management services, offered for enterprises' own customers, for prospecting and for research purposes.

These phenomena we have cited are the results of the application of computer database technologies in the field of marketing. They come from all elements of the marketing conceptualization. To summarize, we believe the implications can be readily perceived as:

- renascent direct salesmanship;
- value-creating speed and availability of information;
- quality prospecting for customers;
- transformed channel development;
- marketing research enhancement and behaviour linkages.

Any one of these areas can be demonstrated as more than sufficient to justify an enterprise taking database marketing seriously even at a tactical level. But in the long term, however, the rewards will go to those who think strategically sooner rather than later. Such strategic thinking must identify the key areas where value will be added for any particular enterprise and then focus on the necessary pattern of organizational change that almost inevitably must follow in its wake.

Little (but there is some) good comes from the enthusiastic recitation of a litany of what computers can do assuming budget and competent staff are limitless. The zeal of the technophile wastes a great deal but such may well be the entry price to the forums of considered debate. The fact that most major database systems so far have regularly failed to deliver hoped for benefits can be the best basis for true managerial concern and commitment next time around.

Beyond the technophiles, we can discern marketing's emerging intelligentsia. Our discussion above of the themes our marketing textbooks must in future reflect are not drawn from our own imaginings but from as thorough an analysis of recently published literature as we have been able to make. We shall now present that literature on the phenomenological basis in which it has arisen.

Conspicuous pioneers
Prospect tracking
Michel[2] reports a comprehensive database tracking system that Cessna developed which managed lead generation and management. It focused on the offer to individuals of flying training linked to aircraft, the end use of which was recreation, personal or business travel. Research among existing customers showed two key prospect groups – those who always meant to learn to fly but did not get around to it, and those who were curious but needed more information. Above the line advertising and direct mail shots to bought-in databases were used to generate enquiries. Extensive pre- and post-testing went into both these areas as well as evaluation of the enquiries generated.

The prospect's first point of contact was indicated in all promotion to be to Cessna itself despite the fact that the eventual training would be done by qualified, non-owned, Cessna Pilot Centres (CPC). The telephones at Cessna were manned by qualified pilots who were wholly knowledgeable about the training programme and the products and gave the initial contact the highest possible feeling of competence. During that conversation the tracking programme required the initial collection and encoding of the area of interest and the source of awareness. The output went immediately both to the field organization nearest to the enquiry address and to the appropriate CPC. The enquirers were, incidentally, verbally notified of the nearest CPC when providing the appropriate details of their address since the database presents the nearest centre's address immediately it was encoded.

To this day, immediately the phone conversation is completed, the Cessna office can either directly contact the CPC or pass the information through on a 48-hour cycle for follow-up. CPCs greatly appreciate the lead generation procedure because it helps build their business but also because it clearly demonstrates the close relationship between the factory and the distributor.

Not surprisingly, there is a detailed feedback loop from the CPC that ensures that no enquiry so carefully generated goes unattended to. The Cessna enquiry database provides opportunities to measure potential promotional cost-effectiveness and scope for future product offerings, regardless of whether the enquirer moves on to join the customer database at once. Most especially those who were clearly curious for more information are followed up in a way that reflects that original enquiry attitude.

Hilty[3] describes a similarly integrated enquiry tracking system for Reliance Electric. He reduced turnaround time by 50 per cent for enquiries from magazine advertising for electrical and mechanical power distribution equipment, scales and telecommunications equipment. His Marketing Information Centre handles over 120,000 enquiries annually and despatches 370 different bulletins to satisfy them. His central function acts immediately to qualify the enquiries while also measuring the differential response from advertising media as at Cessna.

The qualification process is accomplished by the despatch with the designated information packages of a prepaid reply card that offers the option of more information and/or sales visits. It also asks for further in-depth information from the enquirer. While this is being done, field sales managers are notified within 72 hours of receipt of any enquiry, and a Sales Detector form is raised by computer directly for the appropriate sales engineer who must make contact to conclude a sale and/or qualify the lead as worthy of inclusion in the database as a prospect for the future. Total system performance is monitored monthly, and non-returned prepaid card enquiries followed up after 30 days.

More recently, 800 toll-free numbers have been added as an alternative route for enquiries to reach Reliance that reduced turnaround time to the same level as Cessna, i.e. 24 hours as opposed to 17 days for magazine advertisers' coupons.

Qualifying leads

3M Griddle Cleaning Systems evolved a highly cost-effective database marketing strategy[4]. The problem they faced was lack of good quality information on where griddles were located on which their cleaning system could be used. They needed to provide their non-exclusive distributors with high quality leads to follow up if they were to spend their time for 3M rather than working on other more attractive activities for other suppliers of different services and products. They widened their market by mailing to a wide variety of fast food outlets including pubs, canteens and cafeterias, as well as the obviously fast food outlets. A free sample was offered in return for tick-box information. A 12.6 per cent response showed a higher than expected incidence of griddles in non-traditional outlets, details of how they were currently being used, and current suppliers of cleaning equipment.

Cole also described Bowater Scott's lead qualification process for WypAll disposable industrial wipes. A sample was mailed *ab initio* and the feedback of information through tick-boxes was rewarded by a WypAll bench-top dispenser. For such a level of benefit, Bowater Scott sought a considerable amount of information. The 14 per cent response achieved gave salesmen qualified leads by level of usage and concluded with a greatly increased sales call conversion ratio. Of considerable interest will be the fact that the request for information in both cases that Cole cites gained a higher level of response than equivalent mailings without the request. Hesitancy about collecting relevant and appropriate information from interested prospects would appear to have been misplaced.

Gwynn[5] describes a qualification process for raw enquiries generated from exhibitions, news releases and trade shows. He used a "bounce-back" reply card in the fulfilment package and more recently a telephone lead enquiry system, similar to that employed at Reliance Electric. The normal qualification rate settled at one third.

Unlike 3Ms and Bowater Scott reported by Cole, no incentive was included but similar information was requested. Sales were tracked through each enquirer's call record and the cost benefit was established beyond any doubt with average initial sales per respondent to the bounce-back of just under US$10,000 for an incremental qualification cost of US$34.

Important by-products of this approach showed that 99 per cent of respondents were satisfied with the informational content of the fulfilment package and that one quarter of prospects who had not purchased after the time period involved in the process of follow-up were still considering it, i.e. were an appropriate focus for database follow-through in the future. Sales managers were subsequently empowered to ensure their retention in this way until they concluded the prospect was no longer a reasonable one or the sale was made.

Loyalties

Pearson[6] asserts that, by creating a customer database, an entrepreneur is creating the opportunity for a permanent relationship with the customer. That relationship can be loyalty in repeat purchasing similar products and services, in staying with the enterprise at large or with others it might introduce. But it can survive only if sensitively used[7]. A customer database contains a very wide range of different individuals and their loyalty will be given and won for a variety of different reasons. A strategic approach to segmentation and product development is required.

An insurance company typically promoted five year renewable policies to its two million policy-holders. An age overlay linked to product development focused on a four year non-renewable policy for parents with college-aged children. A book club regularly lost 75 per cent of its new members as they became relatively inactive after a year. They received enthusiastic renewal encouragement but, if they had not responded after three reminders, were disregarded. Different media, other than books, were successfully promoted but within the same area of interest.

The issue of greater or lesser levels of activity among customers may well reflect low overall purchasing or a mixed pattern of purchasing. Recency, frequency and monetary value (RFM) analysis of customers can be supplemented by marketing research strategies similar to those described above when prospecting at Bowater Scott and 3Ms. Targeting of promotional activities and new product development can then be clearly focused once those data are available[8].

Time-life Books[9] has for more than a decade cultivated the loyalty of its readers. But by 1983 the marketing manager's overrides of selection rules were becoming a regular occurrence. Modelling created $1/2$ million profit improvement by focusing on "product affinity" across all products purchased rather than individual products. The basis was 100 million mail pieces per year.

Rosenfield[10] described how Carnation addressed the erosion of brand loyalty in the cat food business where their brand is Friskies Buffet. The main source of erosion was discount couponing which although leading to only 4 per cent redemptions, nevertheless conditioned brand choice in the first instance.

Carnation introduced a Friskies Buffet Cat Club with an expiry date for membership providing a renewal opportunity. Repeat purchasing was recorded on the membership card, a magazine on cat care was mailed and discounted cat-related products were made available against proof of purchase. Such an approach was shown to meet the needs of cat owners in an effectively intensive but not an invasive manner.

Savini[11] reported how Helzberg Diamond retail traffic has been developed and built through customer loyalty programmes. No products are ever sold outside the 75 stores. Geodemographic segmentation of what is now a 900,000

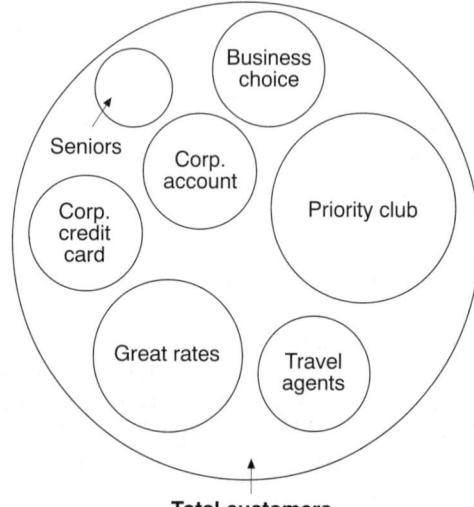

Total customers

Figure 1.
Holiday Inns'
individual constituency
databases built

customer database encourages visits to the nearest store with highly focused promotion activities.

After a decade of such cultivation of loyalties, the evidence is that their "existing customers are at least five times more responsive and valuable than an outside file".

There can be no better note on which to conclude this review of loyalties than to consider Holiday Inns' Priority Club guest register[12]. This guest database has seven important constituents, as shown in Figure 1.

One particular loyalty element is the Priority Club. It began not as Friskies Buffet Cat Club with a comprehensive marketing strategy, but as a points-tracking system and for issue of member statements. In 1987 it was downloaded as a decision support and marketing enhancement function wholly. Its goals were to build a relationship with customers, research the impact of promotions, and to measure the impact of competitive effects to discern how to react. The Priority Club application form has been used to collect relevant information for the customer database, including utilization of competitive hotels.

Already, the database has been used to identify customer life cycles. Eight major patterns have emerged of which one in particular is far more valuable/profitable than the others, made up of not only loyal but high frequency customers. It accounts for 40 per cent of bed nights. They have now entered an enhanced club pattern as Priority Plus, and they get more frequent communication. A further segment, although only 3 per cent of total customers, has an expenditure growth rate of 120 per cent year on year. Together, representing less than 10 per cent of customers, these two groups account for over 50 per cent of total business.

Holiday Inns has taken these analogues of good customers as their overlay for prospecting for new customers. It also used the cluster analysis to identify how vulnerable their programmes were to competitive inroads recently and avoided across the board retaliation which would have been expensive.

Caring and developing
The customer database developed by most consumer durables manufacturers from dishwashers to lawnmowers provides ample opportunity for proactive marketing. Planned maintenance arrangements offer similar scope, not only to sell again but to review the product in use throughout its life cycle.

One of the best documented database programmes in this area comes from Pitney Bowes. Curtis[13] describes his enterprise as tenacious in the use of mailing shots that gather information about potential customers. But the salesforce making 1,000 calls per day, 250 days each year, is also called in to assist. They are expected to audit on all visits any competitive equipment, future prospects for expansion of the company's mailing requirements and who the real decision maker is.

The human element has been harnessed in an interactive database described in Figure 2. All enquiries are speedily transmitted to branches for action as already noted for Cessna and others earlier. But the sales representative also has the opportunity to input ongoing information which can subsequently be monitored centrally.

Most importantly, the database becomes the basis for after-sales tracking and servicing. The Pitney Bowes system is called CARE – call, allocation, review and expediting system. It was primarily designed specifically so that the sales engineer who called would be the most suitable for the customer's problem

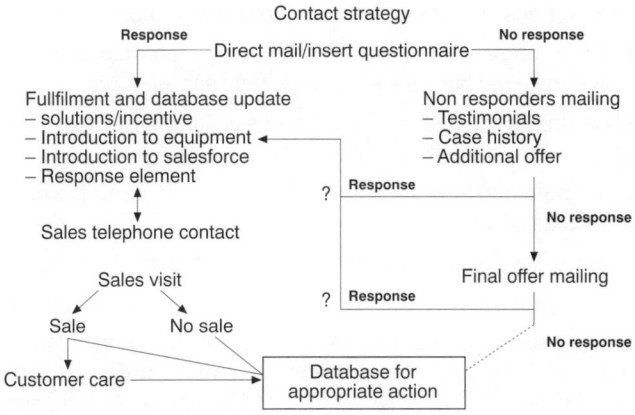

Figure 2.
Harnessing the human element at Pitney Bowes

and have a full record of the installation's service history. Such service histories have, of course, become most valuable information for sales targeting. The database is not only fed from central marketing but "it grows intelligently through real customer contact…It absorbs detailed intelligence about rival suppliers' installations and plans… It enables us to target right down to the individual decision maker who is happy, or unhappy, with our competitors' products".

Two new Pitney Bowes products, the 5030 and the 6100, were specifically developed and targeted to organizations with known mail-room staffing levels and mailing volumes.

Quite to the contrary but equally powerful in the use of databases for product development is Rover's *Catalyst* magazine[14]. Its publisher invited an 800,000 target audience to register in order to receive the magazine and in so doing nominate three out of six life style interests. The magazine goes out with a letter from the local Rover dealer by identified post-code to existing readers and from the Marketing Director of Rover for new readers – giving 67 different runs at print stage. Furthermore, to register for regular copies, the reader must complete a questionnaire about the car they own or with which their company supplies them, and their preferences among Rover cars for the future. They are asked what their replacement car might be. With this information Rover is able to gain a database for test driving invitations.

Will the paradigms shift?
If database marketing can deliver anything approaching the successes described above across a wide field, many of our professional assumptions must be altered. Those who linger too long will fail.

Ditzler[15] opines that:

(1) It may be just the very idea that will shift promotional expenditure away from traditional general advertising methods above the line towards targeting possibilities inherent in direct marketing.

(2) It surely presages the end of marketing research as we have seen it evolve since the late 1930s based on field samples.

(3) It has already transformed the role of salesforces by providing them with qualified leads aided by telemarketing.

(4) It ushers in a new era of comprehensive product and price comparisons at the time of purchase.

(5) It makes possible the effective development of loyalty programmes.

(6) It makes possible the effective conduct of after-sales programmes.

The benefits of such paradigm shifts are so great that sooner rather than later enterprise will seek to embrace them. The outstanding questions must accordingly be: how and in what sequence? The answers would seem to be as old as the history of introducing the first computer into enterprise. The great body of evidence repeats again and again these verities[16]:

(1) Think total marketing architecture but act particularly.

(2) Think revenue streams to prioritize actions.

(3) Piecemeal is bad but incremental is good.

(4) Work pragmatically on issues you already understand and creatively extend your strategies and tactics.

(5) Divide your efforts, 20 per cent on development and 80 per cent on application.

Above all and throughout all, little progress can be expected to be made without top level management support, requiring at least one important advocate. But this can realistically be expected if the five verities are respected, and extravagant promises are not made.

Ditzler[15] focuses on the need to accept that all systems will be tailored to the particular circumstances; and the absolute requirement to agree on definitions at each stage. He goes further into the political and behavioural ownership of the system than most commentators. He insists that the client for the system should supervise data entry to ensure that what lies within will never be dismissed as "garbage".

Nonetheless, Ditzler believes fervently that there must be one common structured file. Customers and well qualified prospects, he asserts, should both be held in common in this way, because as a confirmed optimist he expects all well qualified prospects to become customers sooner rather than later.

To succeed in the marketing arena furthermore, Ditzler is wholly convinced the system must be very user friendly indeed – easy to access, quick in response, menu-driven and with multi-access. Marketers are not computer boffins. They have different information needs from virtually all other managers within the enterprise (except perhaps the purchasing function) in that they look outwards rather than inwards. As such, marketers must become fully aware, as accountants and operations managers have before them, of the potential for computer power. They must develop within the marketing function specialist database experts who work as integral members of the marketing team of the future.

"The marketing database is the foundation for all future marketing", concludes a euphoric Ditzler. "It is the fundamental technology required".

But, as has already been noted, nobody wants or expects everything to happen at once. The pragmatic approach is how the examples we discussed earlier arose and were addressed. Stone and Shaw[17] have traced four phases which are not jumps but evolve into each other. Yet they do have distinctive philosophies.

Phase 1: mystery lists

These are basic databases where little is known of the customer or prospective customer except their demographics and single product purchases. They are normally an outgrowth of accounting procedures or bought-in lists. Duplication is probably widespread and customers' total purchase history is inaccessible.

Phase 2: buyer databases

There is unlikely to be a single marketing database in this phase but marketing and sales planning makes extensive use of databases they have available. Overlapping can accordingly occur to the amused exasperation of customers. The phase is characterized by a determined effort to improve the quality of the data as captured, a broadening of horizons to encompass analogous prospect databases, response handling and fulfilment, and management systems to aid campaign planning and implementation.

Noteworthy elements will include knowledge of the life time value of customers to the company, customer service handling, better closing of the sales loop, and more precise segmentation. Database marketing becomes a critically important element of the total marketing mix.

Dilemmas will still arise if there are several databases as Ditzler argued. There can be disagreement on the definition of a "major account". The salesforce may be deluged with unqualified leads it cannot handle and will normally keep its own counsel, and campaigns can cause inventory run-outs. Stone and Shaw see most companies currently in Phase 2 but learning therein what to do next.

Phase 3: customer-co-ordinated communication

Here one database drives all customer communication and management. The emphasis is not on the databases as such anymore but on the customer. The questions are: Who are our customers? What are their needs? How can we plan and co-ordinate our communication efforts to meet them?

If the database is available to all marketers in the company we may have too many campaigns, and campaign co-ordination is a vitally important element. Campaign management is also required so that all concerned and affected throughout the company may be aware of the implications for them. An expert system of automated decision rules will often be present to guide the management.

Phase 4: integrated marketing

This seems but a short step from Phase 3. It ensures that all within the organization that need to know what, where, how and when marketing is happening get to know automatically and adjust their own systems' performance thereto. To achieve such integration naturally requires that all other systems are at a level where they can thus communicate, which can be achieved only if the enterprise at large has a total information strategy in place.

The critical step, the paradigm shift that begins to make it all happen, is undoubtedly from Phase 1 to 2. Incrementalism cannot make that shift, as Ditzler observes. Political support, top level commitment, is a *sine qua non.*

Thus spake the literature which your authors reviewed. We did so for the excellent reason that since 1977 our own enterprise at MCB University Press has been evolving into database marketing activities. It was our intention to explore what others have said and done to enrich our own approach. As will be detailed below, we did not consciously think total marketing architecture – we worked pragmatically. We did not begin by consciously thinking revenue streams. Lately we were more fascinated by the technology and the sheer common sense of it all. When we reached database marketing proper, we devoted our efforts 75 per cent to development and 25 per cent to application. But in our defence, we always remained pragmatic and courageous, and we tailored our own systems.

Accordingly, when we have provided an analytical description of our own database marketing development, we will attempt our own satisficing model of action learning in respect of the growth and development of the marketing database.

Database evolution at MCB University Press

MCB University Press publishes graduate academic and professional serial titles and associated books and training materials. By 1990 it had more than 70 serial titles and over 160 books and monographs together with a score of training resources packages in print. They are focused to serve five markets, i.e. general managers/senior executives, management development professionals, marketing professionals, librarians and scientists/engineers.

There are some 45,000 active customer accounts with a normal serial subscription list containing some 500-800 names. These accounts have been held as computer files since 1977, initially on a bureau basis but since 1986 on an in-house computer. Books, monographs and training materials are regarded as peripheral to the main subscription activity and will not be discussed further here, though virtually all their customers are also subscribers and very considerable potential for cross-reference marketing exists and has been exploited.

First stage: bureau computerization

The computer files were first created to gain efficiencies and labour productivity in the process of subscription servicing and renewal, which was always seen as the engine-room of the company. However, the director responsible for its initiation was a management statistician rather than an accountant. Accordingly, her personality and intellectual interests very much determined system design. She was particularly concerned to ensure appropriate data were collected to model renewal behaviours across subscriber demographics, including years since first subscribing. From the outset the data captured made it possible to predict renewals with a very high level of accuracy. The computer files were also linked from the outset to distribution/despatch services for all serial issues.

The original system was also greatly influenced by the aggressive marketing and direct promotional orientations of the enterprise. Four of the ten directors at the time were professional marketers and considerable boldness was displayed in direct promotion for most serials. The advertising to sales ratio was typically upwards of 20 per cent which required good monitoring of promotional effectiveness. Ceiling decision rules were set as early as 1978 that the company was prepared to spend 100 per cent of the first year's subscription income on gaining a new individual subscriber and 150 per cent on gaining an organizational or library subscriber. These rules were based on the accurately determined pattern of renewals the data afforded and were very conservative for a well established serial.

The renewal analysis also had several other vitally important marketing roles to fulfil, however. Because each serial title was deliberately directed to graduate academic and professional markets, only small volumes were expected. As such, production expenses as reflected in overall pagination and frequency of issues as well as pricing levels were vital for commercial viability. The renewals model was able to isolate out effects of changes in these. The intelligence that we gained has had a consistent impact on marketing strategy for both existing and acquired serial titles for the last 13 years.

No intensive use was made of the computer files for promotional mailings, leaflets being inserted on a regular basis in all serial issue despatches on a complementary titles basis. The great majority of direct promotional expenditure was devoted to lists of prospects coming from list brokers throughout Europe and America. Media filler advertisements were also placed in MCB University Press's own complementary serial titles but above the line media elsewhere were never found to be cost-effective in comparison with direct mailings.

Finally, although nearly half of all subscriptions were placed and renewed by library subscription agencies on behalf of organizational or library customers all over the world, a single subscriber database was held with demographic details. This was because all despatches for serials (not the case for books and monographs) went direct to the subscriber on an issue by issue basis. This good fortune was offset by the inability readily to trace the source of any promotional influence on the subscriber via an order form because new subscriptions from agents came via their own systems.

Determined efforts were made to encourage subscription agents worldwide to promote MCB University Press serials with preferential terms offered in the first year of up to 100 per cent commission but subscription agents had little understanding of or inclination towards promotional mailing shots whether in Japan, North America or Australia. The only initiative that succeeded was when one director went into residence close by the Australian agent and masterminded successful promotional campaigns from his premises on his behalf.

After considerable exertion with agents in this way, MCB University Press concluded agents were not a vital channel for sales activity – rather a facilitator of subscription renewals. The discounts offered to them were progressively reduced from as high as 25 per cent of all renewals and new subscriptions to a precisely focused fixed service fee conditional on renewal being made not later than 28 February with a retrospective bonus at that date for all subscriptions renewed and settled before 30 November previously.

There were two obvious marketing weaknesses of our first stage computerization. First, the time it took to cross-tabulate subscribers of one serial against another when mergers within the portfolio were contemplated was annoying to directors and others. Secondly, and far more importantly, the customer file from the bureau was at any time up to a month out of date, and customer service, telephone or letter enquiries were increasingly delayed thereby. This was typical throughout the industry but nonetheless a continuing source of discontent.

Second stage: in-house
Between the mid-1970s and the mid-1980s, the economics and capabilities of computers were transformed. At the same time MCB University Press business flourished not least because of the marketing intelligence gained from first stage computerization. Accordingly, the decision was taken to move from external bureau operation to an in-house computer. Although the originating director was still very active and championed the second stage throughout, a systems manager was brought in and devoted a total of two years to systems design, software development and implementation.

As would be expected, it overcame the obvious weaknesses of the first stage systems. Each customer services executive was given a desk top VDU to be able to call up customers' details for updating daily and for answering any enquiries relating to file data immediately on request. Provision was also made for the customer services executives to have VDU access to status reports on serial production and despatch of current issues and for the balance of the volume – a regular request.

The arrangement of customer files and the software development also made possible the speedy retrieval of subscriber list overlaps for direct promotional purposes. This was to become a managerial problem as time went by and more and more marketing services executives and publishers wished to make use of the facility for cross-promotional activity within the customer database. They began disrupting the engine-room process of subscriptions entry and renewal, an unforgivable managerial sin in the organization. Such was the strength of the marketing orientation, however, that this irritation became the genesis of the company's third stage exodus to a stand-alone marketing database.

But that is to run ahead, because a third marketing milestone was passed in the second stage, in many ways the most significant of all, because it was the first wholly integrated database marketing activity attempted. Not surprisingly it focused once again on the core revenue stream of the enterprise, renewals. The campaign, repeated annually, is the Renewals Improvement Campaign (RIC).

RIC's origins lie in the company's ongoing process of action learning management development for its senior managers. The systems manager had reluctantly joined the programme and his topic had been how to improve renewals. It linked with his responsibility for developing the in-house computing system anyway. When he reported back to his fellow managers and the directors at the end of the programme his proposals for improved invoicing routines and chasers were readily accepted. But another strategic theme emerged in the ensuing debate. If the enterprise was prepared to spend up to 100 per cent or even 150 per cent on gaining a new subscriber, surely it should be prepared to spend something towards that level to keep those who already had subscribed. The systems manager's systems-based improved procedures came nowhere near to that potential level of expense. Customer retention should be seen as a competitive weapon[18].

For the first time, the senior marketers among the directors sat down explicitly to seek to model the psychology, even the sociology, of renewal and repeat purchasing as opposed to the statistical results of what was being done implicitly. The model was readily created because of the extensive literature available (much of it previously published in the company's own serials on marketing). It postulates that renewals will be at a maximum if the renewals cycle begins prior to the last issue of a volume. The penultimate issue contains a newsletter to all librarians with details of acquisitions, new journal launches and commentary on the year's activities within the organization. This is followed by the dying issue of the volume which includes literature reinforcing what the serial has accomplished in that volume and a forward look at what the following year is to address. Also included is a renewal notice for non-agent subscribers with a promotional message printed thereon. If not soon renewed, the subscriber is chased and a complimentary copy of the first issue of the next volume is despatched as well. Thereafter, those who have still not renewed are since offered an incentive or a complimentary book/monograph for renewal by a given date and finally telechased.

While the impact of RIC has been difficult to isolate with complete accuracy because other activities were deliberately not halted for experimental purposes, the indicators are a 5 to 8 per cent improvement in renewals since its introduction and a cost per subscriber retained still well below the expenditure ceiling for selling a new subscription. The final telechasing element has broken even on renewals achieved in the first year provided a fresh invoice is raised after the telechase. Simply getting to "yes, will renew" on the telephone underperformed by as much as 50 per cent the eventual renewal compared with the fresh invoice and covering letter. One further refinement that also emerged has been the despatch of the second next volume complimentary issue to non-renewers.

Analysis had constantly shown that the complimentary copy reminder to renew was the strongest element although it seems likely that telechasing now matches it. Again we cannot be definite because of the constituencies of non-renewers to which these two reminders are addressed differ. Telechasing comes last of all. We have pondered why, and are shortly to conduct a controlled experiment that promotes telechasing to an earlier place in the RIC campaign. Needless to say, we are seeking not only renewals but renewals that take place as easily as possible to maximize cash flow benefits (*vide* the agents discount structure referred to earlier), and we also want renewals that are not simply for a single year but continue thereafter.

Another phenomenon has recently also emerged that is receiving careful analysis, i.e. renewals in advance. Recent acquisitions with widely dispersed renewal dates have been experimented with to pull cash flows forward by early invoicing. When incentivization was added in the form of two bonus issues if renewal was received by a given in advance date, 44 per cent did so within two weeks.

While RIC is a most considerable systems and commercial success, it has neglected to treat first time renewals, librarians and organizational renewals, individual renewals, well established subscribers or multiple subscribers as segments. Fundamentally, there is one subscription renewal campaign and everyone gets it or it is left in the hands of subscription agents. That is beginning to change as our ability and awareness that we can manage these segments grow. A further 5 per cent improvement is surely available when we get there.

RIC also triggered another issue of some substance and future importance. Where did RIC belong in the organization structure of the company? It had obviously grown beyond the direct renewal invoicing procedures to include an increasing number of marketing elements.

It sat astride the Customer Services and Marketing Services departments and was also of considerable interest to the five publishers responsible for the profitability of the company's five market missions. One year it fell right down the middle of the two departments when internal issues and staff changes distracted their attention and the directors sent in a task force to recover the situation. The matter has, however, been resolved in the wake of the disaggregation of the Marketing Services department among the publishers and the retention centrally of only a Marketing and Agency Development Unit. Currently, an annual review cycle led by the Marketing and Agency Development Unit, but specifically including all the executive publishers as appropriate, agrees the marketing structure of the upcoming RIC. Thereafter, however, its operational implementation including telechasing is the responsibility of the Customer Services department (see Figure 3). Competitive behaviour to buck the RIC campaign by publishers anxious for their own

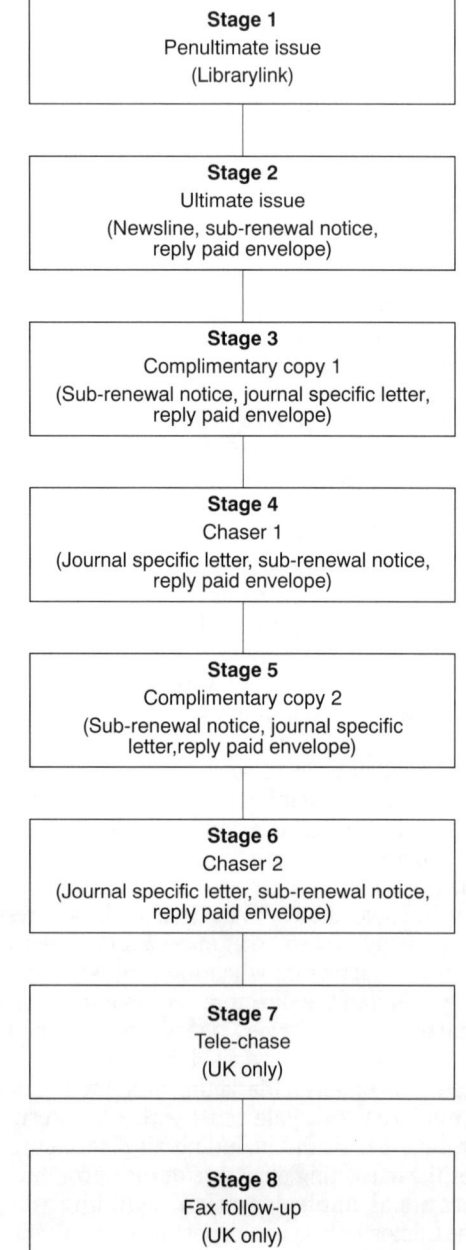

Figure 3.
RIC IV (1991) renewals
improvement campaign

serials is definitely discouraged in the macro interest of the company's overall performance.

The same phenomenon of interdepartmental interface problems arose between Customer Services and the production and despatch status-reporting pattern which the second stage computer facility has the ability to store but not to capture. The need for the information was seldom appreciated within Production and the Distribution Centre. There is, incidentally, virtually no staff movement between these departments. It was not until 1990, four years after stage two went live, that this information finally arrived on a regular basis as originally intended.

Third stage: marketing exodus
MCB University Press's dedicated marketing database went live on 30 November 1989, with a personality of its own called Dick Data (see Figure 4). It was built to marketing's specification and operates on their own departmental hardware. The data which comprise it were originally downloaded from the customer services database and are now updated monthly by further downloads of changes. Campaign data are entered directly by the marketing services executives. Its origins lie in the best tradition of bootlegging intrapreneurship within the now disaggregated Marketing Services department that eventually found a champion and sponsor at director level in the originating director of the first and second stages and the senior marketer. The bootlegging and then the lobbying for a dedicated marketing database grew significantly from the irritation that intelligent use of the new second stage system occasioned. But the head of Marketing Services at the time also had a strong background before he joined the company in database promotions and he found an ally in an analytical marketer who was a refugee from the company's finance department. They, together with the understanding of customer services held by most executive publishers and marketing services executives because they had normally earlier worked there, created an unstoppable momentum in what was always an hospitable environment.

Let it not be overlooked, then, that the marketing exodus was not top-down development, nor was it a computer-inspired activity. It was a hands-on marketing initiative. It pulled the top and the computer experts into a taskforce to get the job done. In the best traditions of MCB University Press towards staff training and development it has been accompanied by a frenzy of staff training in the potential for database marketing and the necessary keyboard skills. The investment pay-back on those activities is unavoidably two or three years. Software development was bought in to a customized, tailored specification.

Dick Data's arrival was awaited with great expectations. Everybody jostled for position to use it in their own particular way and, as in Customer Services since 1986, every Marketing Services executive was enabled to access Dick Data from their own networked PCs. Experience from RIC had shown us just how

Ric IV

Welcome wagon

Gotcha

Big spender

Figure 4.
Several faces of
Dick Data

such rivalries could potentially confuse the customer – not least librarians who were a real prospect for virtually all the serials in print. So, from the outset, the RIC discipline of an annual review by the Marketing and Agency Development

Unit with the executive publishers and marketing services executives, agreed
what for the upcoming year would be the major campaigns among the customer
database.

This way, it is intended to ensure that substantial programmes with
considered measurement can occur. Every use of the customer database within
the enterprise must not only fit within the annual agreed campaign but be
logged with the Database Operations Manager and every pre-prepared stage of
the campaign fully executed and evaluated – as input for the next review cycle.

This determination to see database marketing campaigns through the fullest
planned cycle was nowhere better illustrated than by the manner in which it
revolutionized past systems of sample copy despatches to prospects.
Historically, the enterprise had willingly despatched sample copies to enquirers.
The evidence from complimentary copies at renewal time and fragmentary data
on new subscriptions gained meant it was perceived as worthwhile in all cases.
However, it was not a procedure the company was well organized to handle.
Sample copy request fulfilment was perceived as a clerical venture in the
Marketing Services department although requests often came to Customer
Services. Details were passed to the Marketing Services department and the
sample copy would be despatched but backlogs were normal. One month later
a follow-up letter was sent with a brief questionnaire to identify whether the
requester required further details or if indeed they had subscribed. This was a
bizarre way of tracing whether or not a subscription had been taken, in lieu of
manually checking against Customer Services' entries.

No regular follow-up activity was in place for those who did not respond to
the sample; no second sample copy was sent.

If the system does not sound much like a system to the reader, that would
be a correct assumption. When Dick Data arrived, over 5,000 sample copies
had been despatched but not followed through in the past 18 months. Only the
original manuscript record existed. That this state of affairs prevailed after
more than ten years of a sample despatch strategy was puzzling, for it
constituted the hottest possible prospects file available to the company. But
the explanation of the puzzle was not far to seek. Until the database was
designed to meet marketing needs, it would continue to be simply, even if
vitally, an outgrowth of customer servicing from entry through renewal but
not prospecting.

Since the Exodus, sample copy routines, including the creation of requests
for samples from key customers, and the entry of requests from prospects, have
been given a high profile. So too has the linkage through reminders of
telechasing. The procedures are a virtual mirror image of the RIC campaign
with the starting-point as a prospect who asks for a sample copy rather than a
lapsing subscriber.

Sample copy strategy
Over and above the more professional handling of customers who write of their
own volition, sample copy campaigns are planned. It is the intention to solicit

requests or send unsolicited sample copies to encourage the sales of subscriptions. Therefore the structured sample copy campaign becomes an integral part of every type of campaign. Some campaigns lend themselves more to one method than the other but the opportunities are there to try. To this end we introduced two structured sample copy campaigns with flow diagrams and step-by-step instructions at each stage.

Sample copy strategy (1). There are two arms to this campaign and in order to monitor the effects of solicited versus unsolicited sample copies, a split sample is carried out, i.e. one half of the list is used to solicit requests for sample copies and the other half is sent unsolicited samples (see Figure 5).

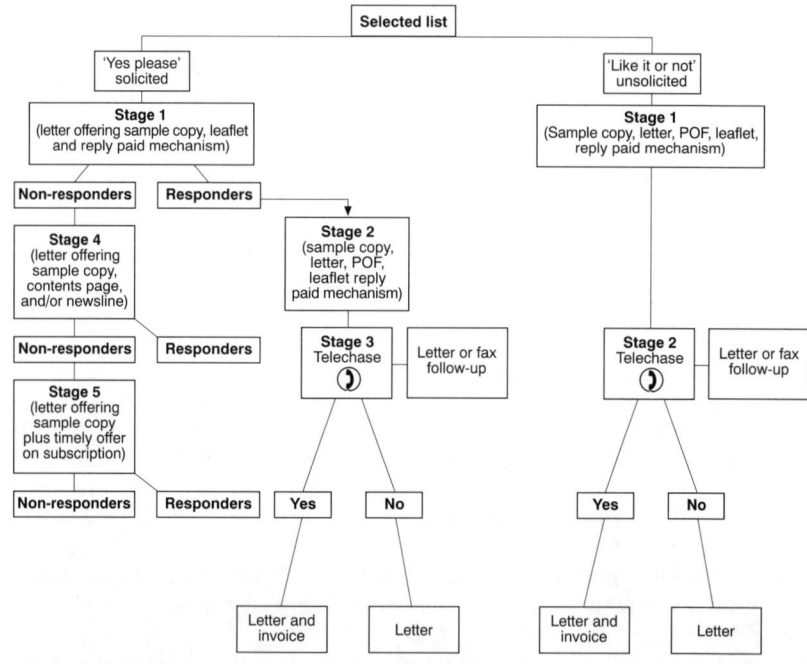

Figure 5.
Sample copy campaign strategy (1)

Note: POF = priority order form

Sample copy strategy (2). Again there are two arms, i.e. solicited and unsolicited, but with a difference (see Figure 6). This type of campaign is used to elicit information on individual managers within an organization about whom we currently know nothing. For example, we may hold information on the Senior Training Executive of XYZ plc but know nothing about the Marketing Director. One thing of which we can be certain is that

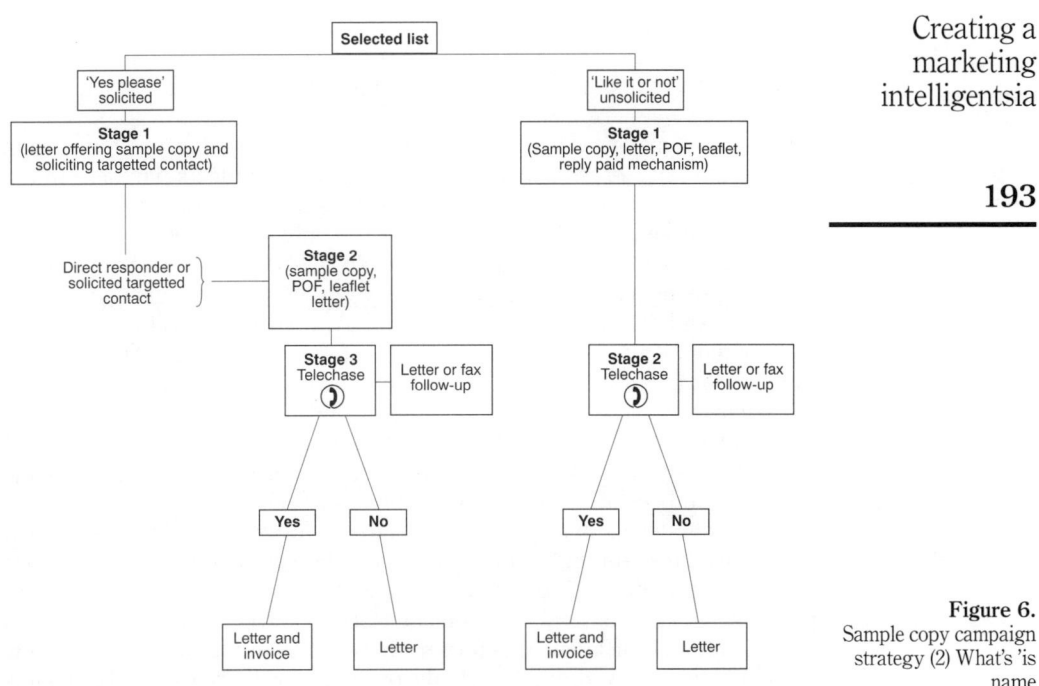

Figure 6.
Sample copy campaign
strategy (2) What's 'is
name

Note: POF = priority order form

XYZ will most certainly have a Marketing Director. So we are basically writing to the person on whom we do have information and asking him/her in turn to pass on details of his/her colleagues to solicit sample copies. Sending an unsolicited sample copy entails that contact being asked to forward the journal on to the relevant person, but informing us of his/her identity at the same time.

Librarians 2,000
In addition to the major drive to thicken up sales with existing customers that swung into action, and the procedures for handling sample copy requests professionally right through to the sale, a strategic use of the marketing database emerged from the Librarians Mission's responsibility to co-ordinate approaches to the library market. A qualified prospect list of 2,000 librarians whose demographics would indicate their potential for MCB University Press serials has been built – Librarians 2,000. It will parallel the models developed with health care patients reported by Kelsey and McGrath[19]. Over a period of two years they will be focused on as sales prospects to be given a sustained marketing and promotional campaign, managed by segment within the database. Taking a return on investment

approach, the equivalent of 150 per cent of the first year subscription for a librarian sale, a not inconsiderable budget, can be expended and amortized over the anticipated life of the subscriptions based on renewal levels gained by RIC.

The fourth major result of the Exodus has been a reawakening of interest in MCB University Press subscription agents worldwide as partners in the marketing process. One of the great weaknesses of Dick Data for growth of database marketing is the fact that there are only 45,000 customers therein. Qualified prospect lists are urgently needed. What better source than our subscription agents who between them have over two million customers in libraries and organizations worldwide? And so it was. The Marketing Development Unit also came to encompass a full-time Agency Development function for the first time in the company's history. Information partnerships with shared data blossomed[20].

MCB University Press is now sharing its marketing database software approaches worldwide with its agents as well as implementing them within the company's offices in Bradford, Singapore and Brisbane. None of the agents have got databases that remotely resemble that which evolved from MCB University Press through its first and second stages and the Exodus. Their databases are directed to the handling of major customers with multiple purchasing needs. Several were reluctant brides at the database altar, but an aggressive use of agent discounts for those who participated for new subscriptions achieved, and latterly the postulated withdrawal of all discounts for non-participation, have achieved a mutually rewarding outcome. Agents worldwide have come to see that MCB University Press is showing them how to grow and develop their own business well beyond the serials MCB University Press publishes. It has become a non-zero sum game for the agents, from which the greatest benefit will flow only when MCB University Press seeks to exploit the agents' databases as they are by further creative development, rather than wishing the agent's database was what it is not. The most likely way this will be achieved is by the same Exodus route as MCB University Press itself used for its own customers in the short and medium term. All other patterns are likely to create too much disruption in the major fields the agent's database was created to service and run into the political rejection area.

Furthermore, until database marketing has had the opportunity to experiment and learn on its own it is unlikely to be able to contribute well and effectively to integrated systems design discussions across the whole field of customer service and marketing within the whole logistics channel.

Fourth stage: information integration
Time is not on Dick Data's side in making progress up the learning curve to understand what can be done with databases in marketing. Customer Services is reaching capacity constraints on its in-house computer and the decision has

now been taken to upgrade it. And from the top down that same statistician and information-searching director has begun to conceptualize and model a total information system for the company. Her objective is to formulate information strategies that take further competitive advantage of what IT offers throughout the major areas of activity in the enterprise. This will include production services and the distribution centre, customer services, financial services, marketing and production, and operate in a way that is compatible or linked with key external participants such as author origination of copy, bankers, agents' information around the world and print suppliers.

It also seems likely that the Exodus may well be short-lived in view of the problems of reconciling each month the download compared with the previous month and separating out the impact of promotion in a convenient way through Customer Services. Dick Data's status in relation to the customer service database is roughly the same as the bureau was at the first stage. His reasonable working life can perhaps be set at no more than three to four years, after which database marketing will again be driven from a centrally integrated computer that is sensitive to the needs of the marketing intelligentsia that Dick and his predecessors have created within the company.

Exploration of the fourth stage has already begun no more than one year after Dick Data went live. The original champion within the Marketing Services department of Dick Data has moved full-time to the Information Strategy taskforce. Left behind in the leadership role of the Marketing and Agency Development Unit, pursuing an 80 per cent application and 20 per cent development mode was the individual most associated with the agency development database. Even this evolution has moved on again with responsibility for the use of Dick Data being placed firmly in the hands of an enhanced executive publisher grouping, and the hardware responsibility passed to Customer Services in anticipation of its shortly to arrive upgraded system.

Satisficing action learning
Most of the steps and issues that Ditzler, and Shaw and Stone[15-17] envisaged have been present throughout the evolution of our database marketing activities at MCB University Press. However, an explanation of the dynamics of the way events unfurled is not provided by them in their papers, so we feel it worthwhile to offer one here against the background of our own descriptive analysis.

The importance of salient issues and personal work preferences must not be underestimated. At the very first stage of development we moved forward on the foundation of revenue flows for renewals as the focus for continuing returns on investment, but without any identification with the accounting function proper. Indeed, the role of accountancy has never been ascendant in computing or databasing owing to the leadership from statisticians in particular. Rather it has been a consistently neglected discipline. The annual planning cycle, introduced formally as long ago as 1967, for instance, has always stayed ahead of the budgeting cycle and been driven by the marketers. This was no accident but a deliberate marketing-initiated strategy, studiously maintained.

The evolving cost/capability calculation of computer technology for the smaller enterprise has also been a major determinant of the rate of progress. During the years the bureau was in use for the first stage, directors were well aware of what was technologically feasible but it was not considered cost-beneficial until the early 1980s. Conversely, the advance from stage two to stage three was determined not by technological change but by capacity constraints brought about by acquisitions policies and the explosion of marketing database thinking at bootlegging, intrapreneurial levels of the enterprise.

Finally, the move from stage three to stage four is an intellectual construct brought about by director involvement in leading edge thinking at an academic level which is a matter of a typicality not yet alluded to in this essay. The directors of MCB University Press throughout its history have been wholly and exclusively drawn from Bradford and Cranfield Management Schools and the International Management Centres. The level of intellectual curiosity *ab initio* and throughout its existence at Board level has been at a maximum.

Yet still, and not surprising, so very much has been missed by the directorate, not least the Exodus stage of development. That arose and was possible, however, because of the contextual framework within which middle and senior managers have always worked with the directors, i.e. as an action learning community. And it is potentially, we believe, the most important element of all in progression to a marketing intelligentsia. Exodus has afforded the opportunity to experiment and learn at great speed in order to return with confidence to the corporate information strategy debate at stage four. Without Exodus we believe, indeed are adamant, the quality of marketing input to stage four would have been as weak as it was with stages one and two.

Without the Exodus, prospecting would not have come of age[21] nor would the extension of qualified prospect thinking to agents have occurred. So we must beg to differ with Ditzler's proposition that database marketing is best part of an integrated system[15]. Yes, but later when it has learned and matured.

Finally, Shaw and Stone's[16, 17] conclusion that phase four is an integrated marketing database stops short of the Gestalt corporate strategy information model we are now developing. The system that encompasses all major information flows and uses them to create not simply a marketing but a business intelligentsia is a realistic possibility. In our enterprise we can already see how that is possible. Management development projects are currently exploring the constituent elements and their interfaces/relationships in the strategic context. The next computer installation will be designed to deliver effective relationships throughout the organization and beyond if Konsynski and McFarlane[20] can be heeded.

We would not wish to conclude these observations, however, by suggesting there are no major obstacles nor by suggesting that we expect to achieve or get close to perfection. We have at all times been reminded of entropy, both in terms of systems and personnel. We have at all times been reminded of the fragility of success in face of staff promotions and replacements for directors, managers and staff. To develop and work with the capabilities computers have today is a thoroughly daunting task that no amount of expert systems can overcome. An atmosphere of intellectual curiosity and continuous learning from action and from the store of knowledge are the necessary handmaidens of the satisfaction needed to go home in the evening relaxed.

References
1. Kestnbaum, R.D., "List segmentation: profit v. privacy", *Direct Marketing*, August 1981.
2. Michel, EM. ,"Cessna's extensive database zeroes in on flying public", *Direct Marketing*, August 1984.
3. Hilty, T.J., "When computers process enquiries", *Business Marketing*, February 1985.
4. Cole, T., "Quality counts above quantity", *Marketing*, 22 May 1986.
5. Gwynn, R.M., "Tracking advertising through to sales", *Business Marketing*, February 1985.
6. Pearson, S. ,"A lasting relationship", *Marketing*, 22 May 1986.
7. Passavant, P., "The strategic database", *Direct Marketing*, May 1990.
8. Wiersema, ED., "Hidden information in your database", *Direct Marketing*, October 1986.
9. Hotchkiss, H.S., "Creating a better continuity model", *Direct Marketing*, March 1987.
10. Rosenfield, J.R., "The trouble with direct marketing is…", *Direct Marketing*, October 1988.
11. Savini, G., "24 karat database", *Direct Marketing*, March 1989.
12. Durr, J.L., "The value of the guest register", *Direct Marketing*, September 1989.
13. Curtis, J., "How an interface database helps Pitney Bowes refine products", *Industrial Marketing Digest*, Vol. 14 No. 3, 1989.
14. Cobb, R., "Beyond courtesy", *Marketing*, 16 February 1989.
15. Ditzler, T., "The future is in database marketing", *Direct Response*, March 1987.
16. Shaw, R. and Stone, M., "Competitive superiority through database marketing", *Long Range Planning*, Vol. 21 No. 5, 1988, pp. 5-24.
17. Stone, M. and Shaw, R., "Database marketing for competitive advantage", *Long Range Planning*, Vol. 20 No. 2, 1987, pp. 12-20.
18. Dawkins, E.M. and Reichheld, E.E, "Customer retention as a competitive weapon", *Directors & Boards*, Summer 1990.
19. Kelsey, R.R. and McGrath, J., "Database marketing targets existing patients", *Healthcare Financial Management*, June 1990.
20. Konsynski, B.R. and McFarlane, E.W., "Information partnerships – shared data, shared scale", *Harvard Business Review*, September/October 1990.
21. Van Doren, D.C. and Stickney, T.A., "How to develop a database for sales leads", *Industrial Marketing Management*, Vol. 19, 1990, pp. 201-8.

Further reading

Baker, B., "Using the trade press to update prospect lists", *Industrial Marketing Digest*, Vol. 15 No. 1, 1990, pp. 27-33.

Bird, D., "How to hit the winning post", *Marketing*, 22 May 1986.

Courtheoux, R.J., "Database techniques: how to tap a key company resource", *Direct Marketing*, August 1984.

Edwards, D., "What's so tough about direct marketing software?", *Direct Response*, June 1985.

Fletcher, K.D., "Information systems in British industry", *Management Decision*, Vol. 21 No. 2, 1983.

Hague, P., "Six companies, six databases", *Industrial Marketing Digest*, Vol. 13 No. 1, 1988.

Lewis, H.G. *et al.*, "Database: is it your creative servant, or your creative master?", *Direct Marketing*, June 1988.

Manmarella, J., "Psyching out list overlays", *Direct Marketing*, February 1986.

Meredith, L., "Developing and using a database marketing system", *Industrial Marketing Management*, Vol. 18, 1989, pp. 245-57.

Morton, J., "How to spot really important prospects", *Business Marketing*, January 1990.

Waid, T., "Taking stock of research", *Marketing*, Vol. 17, March 1983.

Wiersema, ED., "Advanced segmentation's practical parameters", *Direct Marketing*, March 1987.

Chapter 12

Journey to marketing clubland

Thanks to database technology, clubland is transforming segmentation of customers from a proactive to an interactive marketing proposition and qualifying prospects in its wake.

The impresarios of marketing clubland are perhaps the most lavish creation of database technology. Many decline and fall as soon as they arise. Yet when they do succeed they provide some of the most cost-effective routes to marketing success.

Kimberley Clark in the US targets over 75 per cent of all pregnant mothers every year to build a lasting relationship with them. Rentokil's Crown Club relates with most major architects. Lotus Users' Club provides a vital avenue for problem solving and product upgrading. British Airways' St George's Day offer of 50,000 free seats drew five million qualified prospects.

In this article we shall analyse the different styles of clubs that have already been reported on in the literature, and seek to draw tentative conclusions. We shall then present our own experiences as academic publishers in sponsoring three clubs during 1991.

Clubland is customer- not product-oriented

We have been unable to discover any previous attempt at defining what a marketing club is, despite the reality that it is with us today. Dictionaries define social, political and professional clubs but not marketing clubs. As so often is the case on such occasions, therefore, it may be simplest to begin by saying what a marketing club most significantly is not. It is *not* product-oriented.

Branding is product-oriented. Clubland, however, is the marketplace. It encompasses manufacture, wholesale, retail, mail order, telesales, just as branding does, but it is customer-oriented. It has already irretrievably changed the balance of market power and is in the process of shifting the entire marketing paradigm. It has been made possible by database technology. Not only will branding of necessity undergo a major repositioning, but so too will marketing research, new product development approaches, sales prospecting and, last but definitely not least, above and below the line media spending. Clubland encompasses the image component of branding without its product-orientation. Perhaps most importantly of all, clubland pays absolute homage to segmentation, turning it into an interactive rather than a proactive marketing proposition. In its highest expression, the marketing club invites the sponsor's own customers from targeted segments to identify themselves publicly as advocates for the benefits offered and to pay a subscription for the privilege of belonging. In this respect it is a mirror image, surely, of the franchizee's willingness to pay a substantial fee to join a franchisor and to agree to abide by acceptable and obviously appropriate disciplines in that role.

With Julian Wills,
Marketing Intelligence & Planning
Vol. 10 No. 2, 1992

At the bottom of the staircase to clubland, "suspects", non-customers from rented-in lists, await the club touch in blissful ignorance. But, as each club takes form, as customers' characteristics are more clearly discerned, the original suite of products or services that triggered club formation is quickly swamped. The club increasingly defines what its sponsor should offer. From its initial adoption of customer-orientation, customers sooner or later become the dynamic driving force in the sponsor's success.

Marketing clubs defined

It would be too simplistic to suggest that marketing clubs are only those sponsored groupings of customers that have been informed and certified as club members. The formal use of the word "club"is also potentially misleading. Some nominal clubs clearly offer fewer benefits to their members than those enjoyed by informal or even unaware groupings which a sponsor has created. Accordingly, it is more appropriate at this time to define marketing clubs as follows:

> Customer groupings with shared interests sponsored by a supplier of goods or services to build and sustain loyal buying behaviour that accepts the sponsor's need to manage the relationship for profit.

The responsibility for maintaining and developing a club's *raison d'être* rests solely with the sponsor. The successful marketing club is accordingly a voluntary coercive network. As such, suggests Johannisson (1987), it should be analysed not only in accordance with exchange but also with communicative and symbolic roles. These two latter characteristics are increasingly present as customer groups climb the staircase in marketing clubland, pictured in Figure 1.

Surely one of the most widely known and respected marketing clubs must be that founded by American Express. The exchange role within its membership network, the problems of carrying cash, or having one's credit-worthiness assessed, are both overcome for cardholders; and for merchants an excellent factoring of all their customer accounts is provided. The *en passant* benefit of "spend now, pay later" may well either increase the total size of a market or increase the participating merchant's share, or both.

But American Express has carried the club's role a long, long way beyond that described above by deliberately enhancing its merchant's business. It promotes this to its cardholder members in myriad ways. In doing so its communicative role is carefully handled to be of clear benefit to card members as well. Finally, the symbolic role of American Express is continually reinforced in advertising to recruit new members. The card is "not for everyone", but once you are a card member you will enjoy a quality of customer service and personal attention you have "earned".

If American Express as a club has taken two words and given them a symbolic image members relish, others have taken club names that afford immediate symbolism – South Pacific Hotels Frequent Guest Club is known as Kudos. Continental Airlines Frequent Flyers Club is known as Bronze, Silver or Gold Elite – symbolism also of course used by American Express with cards in gold and platinum but exercising great care not to downgrade their regular green.

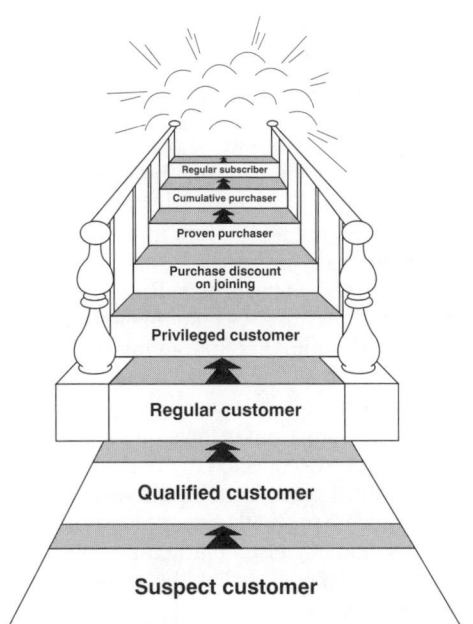

Regular subscriber
Cumulative purchaser
Proven purchaser
Purchase discount
on joining
Privileged customer
Regular customer
Qualified customer
Suspect customer

Figure 1.
Marketing clubland's
staircase

Continental Airlines' Elite Club does not require a regular subscription but it can be joined only by logging sufficient travel miles. The rewards which lie in wait for the member include taking the hassle out of checking in and baggage handling and upgrading to first-class. For reasons of capacity handling, however, few Frequent Flyers Clubs extend the courtesy of a separate lounge for waiting – rather a new subscription membership is required for that. It was for different reasons, therefore, that American Express chose to joint-venture with British Rail to create Pullman Lounges at major termini as an integral membership benefit and thereby build their franchise with first-class rail travellers.

Yet, if the marketing club phenomenon concerns only Stairs 5 and 6, there would be little purpose to this article. They are relevant here only as exemplars of what can occur on lower stairs, even belowstairs among suspects and qualified prospects. The significance of marketing clubland today is that it is spreading across many marketplaces, with considerably lower entry requirements for members. Furthermore, database technology facilitates and customer-orientation dictates that clubland's philosophies and practices are adopted towards all customers, actual or potential, where no formal club has yet been created or even contemplated – and where its investment cost may well be inappropriate.

There can be no doubt, incidentally, that the investment cost of club development and exploitation is a matter for most careful consideration. Club expenses do not pay back on the first offer made to members. No club to our knowledge has yet collected an entrance fee and first year membership subscription that alone meet the expenses of recruitment and induction. Just as with brand-building advertising strategies, club building is a herculean task. It is even more complex and taxing at the design and operational stages because a continuous pattern of personalized relationships must be triggered and managed.

The journey must be made because we know it can be made. It is a journey that will increase the cost-effectiveness of the things marketers do to persuade, influence, sell to customers, care for them afterwards and sustain them in a repeat buying relationship. Its risks of failure in the marketplace are far lower than the approaches they replace, e.g. media advertising to create awareness or direct mail promotion to rented suspect lists. Rather the creative intelligence that arises (Wills *et al.*, 1991) because we take this journey guides new product development, salesforce prospecting and marketing research as well.

A selection from clubland

The literature on marketing clubs from belowstairs to the top of the staircase is wholly case study-based. We have collected together as many case studies as we were able to find and in Table I place them on the various stairs. Belowstairs at –2 we have deliberately cited no particular examples. But rented lists are almost always at the very least somebody else's marketing club. The point to be made here is that, until they have been qualified by the sponsor for their own purpose, they can be addressed only *en masse*.

In an unsophisticated or deadline-driven marketing department, we all too frequently skip internal procedures to try to move our suspects to Stair –1. We put our offers in the mail and simply wait to see what happens. Such behaviour is of course professionally unforgivable and is a major contributory factor in much of the social indignation that junk mail quite tightly provokes. Provided the organization has customer information from higher steps on the staircase, the statistical techniques are readily available to improve performance and to minimize the junk effect drastically. List renters can and should be quizzed with supplemental qualitative and quantitative questions. Psychographic overlays that match existing customers should be used, together with demographics and geographics as appropriate. Wiersema (1987), Li (1987), Courtheoux (1987), Hotchkiss (1987), and Mammarella (1986) all describe procedures that are as applicable on suspect lists as on in-house customers for the qualification of prospects.

Table II records the benefits of limited versus advanced segmentation within a million rented names when either 600,000 were mailed or 760,000 after advanced segmentation.

Qualified prospects

Capulsky and Wolf (1990) cite an outstanding example of prospect qualification. Huggies, the diaper manufacturer in the USA, has invested over $10 million to capture the names of over 75 per cent of expectant mothers on a regular basis. Names are collected from doctors, hospitals and childbirth trainers.

During pregnancy the mothers-to-be receive personalized letters and magazines that relate to the stage of pregnancy reached. When the baby

Stair 6	Regular subscriber	Amex Kudos
Stair 5	Cumulative purchaser	Frequent Flyers Elite Holiday Inns Priority Club
Stair 4	Proven purchase	Kraft Cheese and Macaroni Club Independent Brokers Friskies Buffet Club TEXTAID MUDMAN
Stair 3	Purchase discount to join	Frequent Shoppers Advantage Card Continuity Marketing Part Publishing Gevalia Coffee Rentokil's Crown Club
Stair 2	Privileged customer	Lotus User Base Helzberg Diamonds Tom Thumb Promise Card Ukrop *Sports Illustrated, for Kids* Royal Shakespeare/National Theatre Westminster Residents Card
Stair 1	Regular customers	*DHL Express* for secretaries, *Catalyst, Farm Review,* Bloomingdale's by Mail, UK Arts, Hospital Records, R.J. Reynolds, Philip Morris, Pitney Bowes, Kalamazoo, Sothebys, Derbyshire Building Society
Belowstairs 1	Qualified prospect	Ernst & Young Waverley Press Perkin-Elmer International Paints Huggies Diapers Cessna
Belowstairs 2	Suspect customer	3M Griddle, Bowater Scott, Porsche, Cheerfree Rented lists qualified by the vendor but not by the purchaser

Table I.
Travellers' guide to
marketing clubland

arrives, a coded coupon is delivered with a congratulations message and Huggies can track which mothers try the product. Further technology is being developed to monitor further use. Kimberley Clark, which owns Huggies, has not simply invested in building diaper sales, albeit that per baby consumption amounts to $1,400 annually. The relationship with mothers and mothers-to-be can be leveraged across other affinity products.

International Paints' prospect qualification by its Marine Coating Division required even greater patience than a nine-month pregnancy. Ives and Mason (1990) have reported how they collected data on the condition of all paint on all large ships within their target market. Whenever a targeted ship came into port to be painted anywhere in the world, local International Paints' representatives

	Limited segmentation	With advanced segmentation	
		Goal: fixed circulation	Goal: increased circulation
No. names ranked	N/A	1,000,000	1,000,000
No. names mailed	600,000	600,000	760,000
Response rate (per cent)	0.85	1.08	0.95
No. responders	5,100	6,480	7,220
Gross margin ($)	255,000	324,000	361,000
Acquisition cost ($)	(70,800)	(64,000)	(97,300)
Acquisition cost/responder ($)	(13.88)	(9.87)	(13.69)

Table II.
Advanced vs
limited segmentation

Note: Acquisition cost is gross margin dollars less mailing costs, rental fees, service bureau charges, and the costs of overlay data and segmentation
Source: Wiersema (1987).

logged full details of the vessels, what paint was used and the date. They were able in this way to track the performance both of their own and competitors' products. But, more importantly, it provided shipowners with a comprehensive review of the overall state of their fleets, those which needed repainting and a quantitative analysis showing the benefits of International Paints' own products. The company was able to provide quotations more speedily and considerably increased its market share. The cost-effectiveness of the search for sales was greatly enhanced. The company forecast where it would be rather than waiting to be short-listed.

At a more basic level, Van Doren and Stickney (1990) describe how Ernst & Young and Waverley Press purchased suspect lists from Dun & Bradstreet and Oxbridge Communications respectively. In both cases they qualified the lists acquired and were able to make focused offers to selected enterprises within industry financial services and printing contracting. Ernst & Young was able to network effectively with the information obtained. Waverley Press was able to pinpoint competitors' strong and weak points for attack.

They also reviewed the processes used by Perkin-Elmer to qualify a host of names gathered at conferences and exhibitions for their biotechnical electronic analytical equipment. By geographical and product interest qualifications it was possible to generate a 2 per cent successful response from sales follow-up. Such a scheme, however, seems modest when compared with Cessna's epic efforts reported by Michael (1984). Using a variety of advertising media, tele-reception and salesforce linkages, Cessna were able to perfect a sales-prospecting system for the sale of both private aircraft and flying training.

Proctor and Gamble is reported by Capulsky and Wolf (1990) to have used 800 freephone numbers in its advertising for Cheerfree detergent to target prospects with sensitive skins. Porsche has created a qualified prospect database of 300,000 for its cars. Cole (1986) reports how Bowater Scott used questionnaire research linked to a sample mail-out to qualify prospects for sales visits for its Wyp-All disposable industrial wipes, and he also describes how 3M's Griddle Cleaning Systems used sample mailings to widen the sales canvass in fast food outlets.

Regular customer care
It is a clearly observable phenomenon that many enterprises give greater attention to qualified prospecting for promotion to suspect lists than they do to the development of relationships to sell profitably to their existing regular customers. The explanation is not normally too far to seek. Customer records are kept not by the Marketing or Sales Departments but by accounting systems. Data capture was normally created long before marketers' thoughts turned to database technology. The only sustained contacts would normally be for credit control and customer service. Furthermore, all too frequently the salesforce organization of the enterprise is divisionalized in the field and the total view of customer relationships with the enterprise is missing except in those cases of National Accounts Management, where the customer has normally demanded it in any event.

The state of affairs that Bill Nickoil found when he joined Kalamazoo, as reported by Nash (1988), was exactly that. From the verge of total collapse, the enterprise was turned around by decompartmentalizing the sales effort and cross-selling to targeted customers. The regular customer database became the powerhouse for the campaigns. Sales per customer per year were less than 10 per cent of the purchases in the product fields Kalamazoo covered. Yet the customer base was enormous at nearly 200,000. Using customer account management, with regular sales calls and telesales support, the situation was quickly transformed. Furthermore, the analysis of customers' needs on a regular basis by the salesforce led to the consolidation of all Kalamazoo's software support services into a single house.

Curtis (1989) at Pitney Bowes was not starting from such a low level as Nickoil when he harnessed database technology to his customer relationships. However, it provided linkages between national telesales and mailing campaigns and the salesforce. But it also managed the after-sales trading system called CARE – call, allocation, review and expediting. This level of understanding by Pitney Bowes of its customers always meant that engineering visitors knew what to expect. But, like International Paints cited earlier (Ives and Mason, 1990), Pitney Bowes is fully aware of the inventory of equipment held, its age and performance. Sales intelligence included all equipment whether from Pitney Bowes or another supplier. New product launches could accordingly be readily targeted and in two instances developed as a result of niches that were clearly observable in the field and on the database.

Kelsey and McGrath (1990) report the intelligent use of medical records from patients for follow-up services, with good results despite natural anxieties about scare tactics. Senatar (1991) reports how London theatres are able to use the fact that 85 per cent of tickets are sold by credit card to collect names and addresses and phone numbers to tabulate against the type of arts entertainment preferred. Few efforts have so far been made at regular qualified prospecting, however, on the basis of the techniques available and discussed earlier, so the database is a fairly static constituency, despite some notable one-off exceptions, e.g. the National Theatre's *Venice Preserved* was well supported after promotion to attendees at the Royal Academy's exhibition on Venice.

A number of enterprises have resolved to stay in touch with regular customers via house magazines. Cobb (1989) reports details of British Airports Authority's *Airport*; business travellers will all be familiar with the plethora of airline magazines, of which only British Airways' *Highlife* wins any real acclaim. Mercantile Credit publishes *Choices* to 850,000 customers of a wide variety of retailers. DHL, the freight forwarders and shippers, publishes two magazines, one for each member level of the DMU – *DHL Express* is a tabloid for secretaries; and an A4 glossy is mailed to Chief Executive Officers. All these are designed to enhance the relationship with the customer as a whole, not simply to sell a particular product.

Farm Review in the US and *Catalyst* from Rover in the UK provide segmented issues that differentiate between customers' concerns for pig or dairy farming with an incredible 9,000 variants per month; and for the timing of new car purchase/test drives based on the last purchased vehicle and declared replacement cycles. Such profiling within already segmented customer groupings is very much a future trend, as the cost of achieving it tumbles technologically – both for the matching of customer information from the databases and the printing of the several segmented issues.

Bloomingdale's by Mail has its story told by Ostrager (1986). Its particular fascination is that it utilized advanced segmentation analysis on its own transaction files. A wide range of highly effective cross-selling promotions was developed – ready-to-wear clothing customers were more willing to purchase shoes by mail and vice versa. The major conclusion, however, was that Bloomingdale's million plus customers could be collapsed into effectively three mutually exclusive merchandize categories, namely:

- *Classification 1* – ready-to-wear menswear and fashion accessories;
- *Classification 2* – cosmetics, gifts and table-tops;
- *Classification 3* – home furnishings, day wear and childrenswear.

While with hindsight some considerable logic can be discerned for these three classifications, without the analysis the decision to subdivide the catalogue into three for all 300 selling departments would scarcely have been taken, nor the decision to identify what were called 26 "buckets" or clusters among the 300 departments where cross-selling opportunities were at a maximum.

Taylor and Oake (1990) undertook an equally impressive analysis of 450,000 customer accounts from the Derbyshire Building Society at its 60 branches, enhancing the quality of understanding of its customers with BMRB's Target Group Index as an overlay. This in turn predicted directions for future development, promotion and growth.

Our review of regular customer care can be usefully concluded with two unusual examples. Rapp (1989b) describes the battle fought out during the 1980s in the USA and elsewhere between R.J. Reynolds and Philip Morris. It began when R.J. Reynolds mailed 80 million questionnaires on consumer behaviour, specifically focusing on tobacco but also encompassing other consumer goods. Brand promotion and conversion from competitive brands became a critical focus. What motivated R.J. Reynolds, then Philip Morris, to act was the pending threat of advertising bans for their product category. What they have established *en passant* for the future is likely to be an oligopoly position, such as Nielsen enjoyed for so many years in retail audits. Nor must their significance as a countervailing power to the growth retail customer databases be overlooked.

Such databases are sweeping away several but not all the roles of above the line media advertising, as was foretold at the start of this article.

The second unusual example is from Berman (1987) and describes the use of database technology for catalogue promotion at Sothebys. As with Bloomingdale's, it is concerned with both the supply side and demand side of marketing. The procurement and logging of art items for auction, the creation of a customer-segmented catalogue rather than being necessarily product-focused, and their mailing to those most likely to be interested, have greatly improved performance among their 500,000 clients. It has also been possible to increase the pattern of search on a segmented basis for art items in areas where a good collection is beginning to build by contacting those who are most likely to want to send items for auction.

Privileged customers

Whereas the step from qualified prospects to regular customer care is reasonably distinct, it must be conceded at the outset that the allocations made on Stairs −1 to +6 are judgemental, not to say somewhat arbitrary. A more robust taxonomy will no doubt emerge in the future. For our purposes, however, we do think regular and privileged customers can usefully be separated. The major platform pursued is discounting without membership of a formalized club.

Beadle (1988) gives an excellent example in his discussions of the care, feeding and development of the Lotus database. At the outset in 1986, his organization had separate databases for users, sales leads, seminar attendees, dealers and corporate contacts. Customers were either loved to death or ignored. The foundation of the integrated database he developed was the individual and company name, and the address. The high level of staff turnover in computing made it advisable to focus on the company name as well as the individual. New user registration cards, the major way to capture customer names, had just a 15 per cent return rate. It was heavily simplified, made unmissable when opening the product, and if returned gained "Your Special Privileged Customer Lotus VIP Card – free of charge". It gave entitlement to certain free support services and discounts. Registration rates immediately rose to 30 per cent but Beadle also wanted to capture past, unregistered customers as well. All advertisements and out-of-house mailings invited such individuals to register and gain the privileged status available. All hot line enquiries were also captured in this way. In two years the initially deduped list of 18,000 users had grown to 80,000 including qualified enquirers. He had created the UK software industry's largest database, notwithstanding Kalamazoo's success reported earlier.

Beadle advocates constant use, on a monthly basis, to build a relationship, since Lotus virtually never meets its customers. They are all dealt with through 25 "superdealers" and 900 other dealers. The bi-monthly *Lotus in View* Newsletter has practical tips, questionnaires and invitations to conferences. Such information is not distributed solely to Lotus users. Enquirers are tempted on a regular basis with news of what they are missing. Lotus office staff are regaled weekly with details of successful promotions and invitations to participate.

The database is the ideal platform for privileged offers for equipment upgrades. A mail shot to customers without the graphics package recently pulled a 20 per cent enquiry response, 10 per cent within the first four days. Last year's average success to mailings of the database was 7 per cent. All leads are passed, as at Cessna, to the dealers but a careful tracking is made. Like Pitney Bowes, Lotus's regular questionnaire surveys have established its own market share and on what machines its software is used.

What upgrades are for Lotus, gift occasions are for Helzberg's Diamonds. Savini (1989) describes how during the 1980s Helzberg built a database of 900,000 customers for its 75 retail outlets, analysing their recency, frequency and value as well as zip code location, which in turn permitted a range of psychographic overlays. Cash customers were targeted for Helzberg's charge card with an opening $50 voucher to spend on the first purchase above $100; but no mail order activity was created. "Celebrate mailings" for Mother's Day, Christmas and anniversaries were incentivized.

The Royal Shakespeare and National Theatres have also both created privileged offers for their "Friends", most importantly early bookings linked with discounts. Westminster City Council, Senatar (1991) records, has gone so far as offering all its 200,000 permanent residents a ResCard at reduced prices and 23,000 have taken it up.

Sports Illustrated for Kids (Kennedy, 1991b) has meticulously collected the names of US youngsters under 14 and their parents, from purchase records, and has been able to act as a vehicle for privileged customer marketing in the family DMU. Eighty-three per cent of parents read it, 43 per cent in conjunction with their children.

Two final examples in the USA are the Tom Thumb Page Promise Card (Ives and Mason, 1990) and Ukrop (Raphel, 1990). Tom Thumb Page shoppers electing to enrol receive a monthly newsletter describing their entitlements to discounted items. When they visit the store the Promise Card is swiped at the cash register and all discounts are automatically applied. Demographics, etc. from the application form are readily available for analysis against products purchased and Tom Thumb Page offers manufacturers the opportunity to provide merchandize for inclusion in the monthly newsletters. Ukrop employs similar procedures based on check-out barcode swipes.

Purchase discounting to join

There is a strong school of thought that suggests customers are best using their own initiative to purchase goods and services before marketers invite them to join a club. Otherwise, the supplier will waste much time and energy without any serious future potential. However, it is clear in part publishing and other forms of continuity marketing that the purchase discount to join can give excellent results. Grossman (1990) sums up the issue: "Should a continuity marketer lean heavily on selling the initial shipment or try to sell the whole series?" Despite a prodigious number of tests, it remains a balancing act with no definitive answer.

A successful discount-to-join offer is that provided by the McCall Cooking School. The upfront offer could scarcely be stronger:

> ...send my Free Gift – three sets of 12 Recipages each, two issues of *McCall's Cookery*, a loose-leaf binder and 18 Index Dividers – along with my first set of 12 Recipages. If I decide to continue, bill me just $4.95 plus shipping and handling.

All such continuity marketers do, of course, have to pay particular attention to the initial offer title. In the fist batch of recipes the one that is most conducive to repeat purchase must be spotted. Giving the customer a choice of initial order on the original order form can be a very effective way of improving continuing responses.

Time Life Books is one of the most successful and enduring part publishers. One of the most fascinating elements of its Club World is that any customer who stays to the very end, for example a full four and a half years with 39 full payments amounting to $600 for the World War II series, receives a digital alarm clock gratis on conclusion plus the first book free for any of ten subsequent series.

It would be wholly incorrect, however, to give the impression that continuity marketing is limited to part publishing. General Foods has been a leader in some unusual fields – most particularly Gevalia Coffee from Sweden, which is shipped on a regular basis. Wallace International has built a worldwide reputation in creating store traffic with cumulative offers of cutlery, table-ware and glasses. The trading stamp companies of the 1970s did likewise, as did the Co-operative movement's dividend and stamps for well nigh 150 years. It is perhaps appropriate to observe that, in the USA today, the household penetration achieved at its height in the UK by the Co-op movement is equalled by the 40 million Shopper Advantage Cards in circulation (Rapp, 1989a).

We can conclude these observations with one of the more extraordinary clubs developed and reported by Whitehall (1989) for Rentokil. They resolved to establish an affinity club focused on surveyors, building society managers, estate agents and other property professionals. Named the Crown Club, Rentokil offers first-year complimentary membership and looks for subscription payments thereafter. Members, of course, provide personal and corporate details on joining and a regular bi-monthly magazine *Arena* is despatched. It includes articles on property care but also on events and makes special offers to members from Dunhills and Burberry. Travel, car hire, hotels and health insurance are included as well. Membership brings rights of access to Rentokil's R&D facilities, training courses and seminars.

Proven purchase

Anxieties that an initiative like the Crown Club may be "all very well but will it lift profits?" start to disappear in clubland when proven purchase is required as the entry criterion. The Friskies Buffet Cat Club, reported by Rosenfield (1988), was introduced to seek to reinforce loyalty to the brand at a time when couponing was undermining it badly and driving the product grouping to a commodity status.

Membership was symbolic, i.e. free – a card was issued and required annual renewal. Club members received a regular magazine with advice on cat issues, rewards for regular purchase by way of discounts on related products (not cat food) against proof of cat food purchases. Its determined attempt at "deconditioning" coupon users away from their dependence is noteworthy, although it is more widely practiced in catalogue marketing and insurance to wean customers away from price shopping by the determined adding of value to services.

When Carnation sold its UK petfood business to Nestlé, Miles (1991) describes how a five million name database built from door-to-door sampling was turned into a publishing venture for *Tails* rather than a club. Market research showed only 10 per cent felt a conscious club worthwhile. Benefits so far are seen to be a soft sell, as are those achieved with Kraft's own badged Cheese and Macaroni Club for the kids, reported by Capulsky and Wolf (1990).

Soft, however, is not the word to describe New Scotland Insurance's Independent Club (Rawstorne, 1991) established when the new owners acquired this general insurance business from Allstate. After reducing total agencies from 10,000 to 3,200, 150 were selected for focused attention at the highest possible level. They were given greatly increased commissions and profit sharing. A 24-hour special support service was introduced together with field visits. Free training was provided for the staff and marketing department support in all direct marketing campaigns. A regular newsletter, senior management get togethers and week-end entertainments were also provided. Premiums rose in the first year from £2 million to £11.2 million. The claims ratio among Independent Club members' business settled at 37.8 per cent compared with the company average of 47.9 per cent.

The success of this initiative in general insurance led Scotland to launch a second scheme, Merit, for brokers specializing in home and motor insurance with 500 initial members. They got unique, tightly segmented products with a guaranteed speed of settlement so necessary for domestic insurances. Sales doubled in the first year and retained business rose from 70 to 80 per cent.

Two similarly impressive instances of support for proven purchase are provided by Ives and Mason (1990). The Missouri Book Service (MBS), with just four college bookstores, developed computer software known as TEXTAID to handle the availability of second-hand texts, the prices to be paid for them and their eventual resale prices. The system was offered free of charge to other book stores if they agreed to let MBS have a guaranteed minimum number of texts each year. It was thereby able to position itself as a clearing-house for second-hand texts and subsequently for remaindered stocks of certain new texts.

Baroid was similarly able to create its MUDMAN Club based on expert systems. Whereas Rentokil offered R&D advices and a hot line, MUDMAN offered answers on the most appropriate drilling lubricant to use in given temperatures, chemical and geological conditions.

Cumulative purchaser
The confidence to insist that only cumulative purchase levels bring club rewards is at the opposite end of the spectrum to discount privileges simply for joining. Yet airlines and hotel groups have no difficulty seeing that to be their strategy. Airline Frequent Flyer programmes are directed at gaining the loyalty and custom of business travellers. The more miles travelled, the better the reward will eventually be 100,000 miles travelled can bring two gratis first-class tickets across the Pacific or the Atlantic.

There is better documentation in the literature, however, of hotel cumulative purchaser programmes, most particularly that offered by Holiday Inns and reported by Durr (1989). Holiday Inns' Priority Club has been running for nearly a decade now. They report that they have created a customer loyalty which was previously nonexistent. Hotels are selected by location not by brand – albeit that there were some negatives which occasioned avoidance. The airlines say the same about business travellers who are now Club-loyal, whereas before it was simply schedule convenience. Discriminant analysis of the Priority Club members' behaviours showed eight distinct segments. Cluster 4 (constant usage pattern) and cluster 8 (rapid rise in purchases) were perhaps most noteworthy, although analysts feel there was a need to understand why other clusters were in decline. If overall there was a detectable life cycle spanning five years for club members of rise, stability and decline, how could each phase be managed? Incentivization for the new member may well be resort incentives, whereas later on special recognition benefits may be more appropriate, such as executive floors and breakfast facilities. Airlines have found hassle-free check-ins and lounges to be a major benefit for first class travellers.

Emerging conclusions
The case histories described here indicate a rich variety in marketing clubland, but few clear theoretical propositions. Many nominal clubs have less apparent profit potential available to them than dedicated efforts promise at regular or privileged customer levels. There appears to be no logic in

asking for an entrance fee, delaying it a year or offering discounts, just to get a database record.

Among all these solutions, therefore, what are the underlying purposes?:

(1) Husani (1987) puts his finger on one of the major issues, which has been an increasing defeatism among Chief Marketing Executives about their ability to sustain systems of lead generation among qualified prospects. Time and again ambitious schemes fail. But, for whatever reason that may be, the emphasis has to be on cultivating existing customers. They will respond better, as well authenticated evidence in countless situations shows response levels to be up to ten times better than suspect lists.

(2) Clubs can be perceived as a mechanism to protect market share among customers. Roscitt and Parket (1988) describe it as developing "Proprietary Persuasion Processes" that only the provider of the products or services in question can offer, creating a new psychological monopoly that echoes what branding provided in the 1950s, 1960s and 1970s. Defence of market share built on information not available to competitors is likely to be far more effective. They summarize their views in Figure 2.

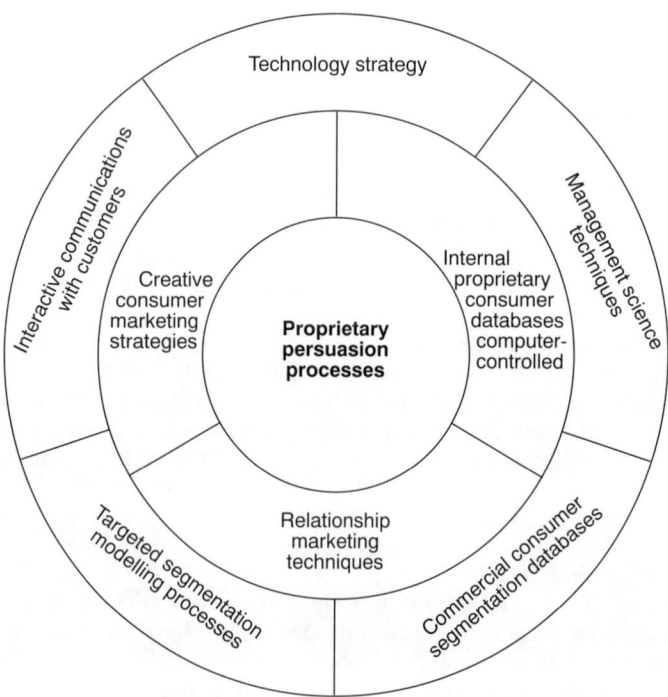

Figure 2.
Proprietary persuasion processes

(3) Clubs can be seen as a vehicle for marketing research in general but particularly pricing and new product development and launching. As Kennedy (1991a) observes, if your own regular customers will not buy a new product, who will?

The depth of information previously bought in from consumer and user panel agencies is now available on a far more cost-effective basis. An approach which provides such a basis for more intelligent marketing must be of most considerable value. Craig (1990) goes further. He asserts that extant customers in industrial markets can often be found to have already developed ways of adding value to the products and services provided. Joint venture opportunities may emerge from complementary vendors.

(4) Clubs can be used as a powerful database for analysing life cycle behaviour, as Holiday Inns' Priority Club has clearly indicated, thereby providing the basis for discounting those who are more and those who are less likely to respond to promotional activities (Chambers and Fleischer, 1986). It goes beyond simple recency, frequency and value propositions to what Passavant (1990) calls the "whole" customer.

(5) Clubs permit the customer to belong and identify personally with the commercial activities involved. Clubs can enhance the felt status of the individual customer and give reassurance in sometimes uncertain situations. American Express is master of this in many of its promotional campaigns above the line. "'Tis pleasant to see one's name in print", Pearson (1986) reminds us. It is also most pleasant to be thanked for our custom personally, even rewarded for it. Raphel (1989) indicates that a range of such personalized approaches is achievable within a marketing club framework that always seems to work.

(6) Clubs are seen as the future for reasons of promotional efficiency and technological productivity. Ditzler (1987) and McKenna (1991) have no doubt whatsoever. Any marketer who does not participate will be at a competitive disadvantage. "Once the sale has been made, why should we start all over again?", asks Ditzler. He concludes by bringing us full circle to the first of our conclusions as well. Once we know who our customers are, the task of qualifying prospects is far more feasible. Furthermore, the existence of club status within an enterprise gives continuing momentum to and concern for the process of moving suspects to qualified prospects and then within the club framework proper. The fact that such strategies are difficult to implement is, as Giles (1991) reminds us, the normal state of affairs. His answer, to devote more attention to ownership in the name of marketing kinetics, we have discussed elsewhere (Wills, 1991).

Developing marketing clubs at MCB University Press

For all the conclusions identified in the list below, the logic for club development at MCB University Press was irresistible:

- as an alternative to the waste and lack of follow through in prospecting;
- to protect market share;
- as a vehicle for product research and development;
- as a marketing research vehicle for life cycle analysis;
- to enable customers to belong;
- to achieve productivity gains in promotional activities and to show how prospecting can be well done and drawn forward to enhance the Club.

We have described elsewhere (Wills *et al.,* 1991) how the enterprise's approach to direct marketing and database technology had evolved to the point where establishing and managing such clubs would be feasible.

It is a characteristic of academic journal publishing, of which MCB University Press is the largest specialist English language management provider, to have small subscriber bases for each title but for each subscription to be of high value – ranging from £80 to £600 per annum. Until the initiation of marketing and sales planning (MARSELPLAN) in 1986, and an independent database in 1990, the focus of promotional efforts had been by journal. All data were held by journal title. While we had always realized that several of our customers took more than one journal title, it was a laborious exercise to find out exactly who. Special computer runs were arranged whenever we envisaged merging two titles.

Apart from that time of year when annual subscriptions were due for renewal, no communications other than the regular issue of the journal were personalized for customers. Cross-selling promotions were done on a non-deduped basis, where titles were obviously complementary, e.g. among marketing professionals, librarians or human resource professionals.

Such renewal activities and promotions were cost-effective and the organization was also able to resource a sufficient programme of acquisitions and new launches. It did, however, miss out on a wide range of productivity improvements that MARSELPLAN brought into sharp focus. It identified the company's two major customer groupings to be:

(1) librarians as brokers for library users;

(2) human resources professionals purchasing on their own behalf.

Even with the database represented on a customer basis, we would still know very little about customers other than how to deduce promotional mailings in cross-selling. We accordingly launched two marketing clubs, one for each major customer grouping. They are known as:

- Library Link;
- Human Resources Network.

Both resolved to approach first those customers who took two or more journals. This gave just over 2,000 librarians and just short of 1,000 human resource professionals. In doing so we made a big assumption, of course. We assumed that the higher value the customer, the more likely they were to appreciate the benefits of the club. This was a big assumption we are currently reconsidering most carefully, to take obvious account of recency but also in the light of emerging demographic information from club members thus far. Both clubs began with a questionnaire approach. Library Link's was four pages long, seeking information on a range of issues such as library size and budget, the dynamics of journal subscriptions, and other areas of interest covered by MCB University Press but where no subscriptions were currently taken. The first recruitment drive took place during July/August and a draw for a £1,000 travel and conference bursary was offered for all replies received by 30 September. After 12 weeks a 20 per cent response had been gained and telemarketing was adding to the membership daily. A 1,500 target response was set. Membership of Library Link was offered free to customers if they returned the questionnaire, and the range of benefits is set out in Table III. Questionnaires were also designed to elicit the names of fellow librarians and the staff involved in the purchase and repurchase decision for journal titles.

We had no idea what response level to expect beyond that indicated in the literature – 20 per cent and rising seems to be satisfactory. The major problems encountered seem to have been mailing at the "wrong time of year" for the Northern Hemisphere Libraries. They were often on vacation. Second, only 10 per cent of librarians were known by name. Whereas a library receiving a journal issue knows how to process it, a library receiving a questionnaire might not necessarily know where to pass it for effective completion. Finally, the travel and conference bursary proved not to be so enticing as we imagined it might be for those who did respond. Fully one half of them declined to tick the box that would have allowed them to participate in the draw on 30 September.

What did come back, however, at once provided the basis for two-for-one campaigns in deduped areas with a high response level received. Such immediate promotional activity campaigns were always a strategic objective. Without the reinforcement of the activity gained from success in promotions, it would have been inappropriate to look for approval for further investment in the database.

It is interesting to note here that, when the Library Link Club was formally constituted, its two major benefits, listings of pre-publication contents pages and its *Newsletter,* were already in place on a desultory basis. Both had been launched as "good ideas" three and five years previously respectively, but their existence was peripheral and their despatch, often a chore, had got forgotten or frequently done too late to be of any great benefit.

The second club, the Human Resource Network, had none of its key benefits in place at the time it was launched. A *Newsletter* had appeared in spring 1989 but only ever got to issue Number 1 before it was forgotten. Nevertheless, it has outperformed Library Link by 125 per cent. It offered free charter membership in year 1 (£50 thereafter) and despatched a short and snappy questionnaire. Replies flooded in from nearly 60 per cent of customers contacted. The balance of information that Library Link sought *ab initio* will be sought later in the year. We have pondered long and hard whether it was the questionnaire length or the wholly different nature of a librarian as a broker and a human resource professional acting on his own behalf. Whatever the answer, far more customers have joined as a proportion of the total customer database and two-for-one offers developed have generated pleasing response levels. The benefit statement for the Human Resources Network is set out in Table IV.

Table III.
Library Link

What FREE Charter Membership
Will Do for You

● Every month you will receive advance notice of forthcoming articles in journals pertinent to your field. This will enable you to take the correct action and make sure the right people in your organization get to see the information.

● You will have direct telephone access to the Network Co-ordinator, who will be able to answer your questions, and refer you to articles in your field.

● You will have the opportunity to submit your own articles and papers for publication in MCB's Human Resource journals.

● You will receive personal priority treatment from MCB when ordering journals, requesting reprints of pertinent articles, enquiring about sample copies, etc.

● There will be regular free trial offers on selected Human Resource titles in your area of specialization. Plus you receive pre-release information of new publications and initiatives.

● Network members will be offered discount prices on Human Resource books.

● You will receive a FREE subscription to the new *Network Newsletter,* which includes up-to-date information on current topics, conference news, case studies, details of training events and courses, as well as letters from other Network members, prize draws, competitions and crosswords.

Table IV.
The Human Resource
Network

The lessons learned at Lotus (Beadle, 1988) were not lost on us in terms of getting staff throughout the company onside and owning their clubs. Bruce and Duncan (1990) and Duncan and Hedges (1991) had already amply demonstrated the difficulties with the broad-scale marketing database. Accordingly, a competition was run with its own prizes among all staff for incentive ideas to be used within the club frameworks. A reception to toast the winners was then held.

MCB University Press' World Sales Conference all the signs were distinctly encouraging. The club concept looks extremely workable, provided that:

(1) a larger number of members joined to provide the minimum critical mass to justify the overheads committed and necessary to maintain momentum; and

(2) the technological/database problems created were overcome.

The second was strictly a technical problem and has now been addressed by the third club development outlined later. The challenge to recruit a sufficient number of members to each club, however, can be solved in two ways – either to collect less information or widen the canvas to include all customers in the club areas of interest. Once the major customers have been given first refusal, so to speak, it seems appropriate and necessary to reach out to all customers next. And, after that, the reach can surely go further to qualified prospects, whose characteristics have been derived from internal segmentation of the full club database. Encouragement has already been received in this respect when looking at the recency criterion, where 5 per cent willingness of new solo subscribers to human resource titles to take a second title and thereby qualify gratis for Network charter membership has been achieved.

At the same time as this issue is being addressed, the Regional Offices worldwide will be invited to take up a proactive role in club recruitment. For the launch phase, all activity was master-minded from the UK head office in Bradford and the United Kingdom achieved the best response levels. However, two thirds of potential club members are located outside the UK and it can be confidently hoped that Regional Offices will do better from Alabama, Singapore, Tokyo and Brisbane.

The Literati Club
Lateral thinking at MCB University Press has caught the club fashion. One of the eternal problems of academic journals is a regular flow of excellent articles to be refereed, revised and published. Until the mid-1980s, it was assumed good editors and editorial advisory boards would do a good job and almost all did. However, with the evolution of production technologies it became apparent that a by-product of the production process could be an Authors' Club (Breen et al., 1991; Wills, 1992). The names, addresses, areas of interest and keywords created for contributions published could be used to provide excellent intelligence for calls for papers. Furthermore, ever mindful of the need to recover the costs involved, authors' own home-based librarians could be expected to be interested in what their colleagues were publishing. The new

service launched in 1992, and known as the Literati Club, can now offer considerable assistance to editors in their search role and provide opportunities to develop library relations, possibly leading to Library Link membership.

The retrospective search through data files back to 1989 to establish the club is one of the best demonstrations we have come across of the need for data integrity. Nonetheless, by the end of 1991 the Literati Club has a membership of 3,500 and will grow annually at a rate of 1,500. Membership is already being surveyed to establish authors' willingness to participate in disk formatting for article submission and to recommend colleagues engaged in research as possible authors.

The benefit statement for Literati Club is given as Table V. Blenkhorn and Banting (1991) describe it as reverse marketing – the application of marketing strategies within the supply channel.

Learning our proprietary persuasion processes

It is already clear that the general principles that were discernible in the review of marketing clubland can be valuable indeed to MCB University Press in its own clubs. However, there is no template that can be readily adopted. Our Proprietary Persuasion Processes, as identified in Figure 2, are in the process of emerging:

(1) Without realizing it and certainly not exploiting it for marketing purposes, each journal and subscriber list stands in its own right as a club. Loyalty is to the journal not to MCB University Press as its publisher. We have made little use in recent years of that relationship but must certainly consider how to do so in future. We have been mesmerized by the initial sale and the renewal decision to the point of exclusion of what else might be done in between which of itself could enhance the renewal response rate that already stands between 70 per cent and 90 per cent.

(2) No life cycle analysis has been undertaken on subscribers as Holiday Inns has accomplished, but it can begin.

(3) Despite an excellent and long-standing statistical tradition in MCB University Press, it has mainly been focused on the beta-geometric model of repeat buying behaviour for budgetary purposes rather than on segmentation to qualify prospects.

(4) Early in 1992 we will be for the first time insisting that high quality, high response promotions to club members and to the host of other regular subscribers are implemented first as a top priority and only then followed up with suspect/qualified prospect listings. Thus far all marketing clubs are the residual legatee of promotion rather than the doyen.

(5) A process of "deconditioning in-house copywriting from emphasis on the price dimension of the offer" to focus on added value through service and product enhancement instead may also now begin. Copywriting and

LITERATI
· C L U B ·

Literati Club is an *exclusive* service to MCB University Press Authors, Editors and Advisory Board members commencing 1 January 1992

Your Literati Club offers the following benefits

Future publications
Priority given to all articles you submit yourself or you wish to commend.

Photocopying rights
You are authorized to make up to 25 copies of any single article published by MCB University Press, without seeking prior permission provided they are not for resale.

Calls for papers
All editors' calls for papers in your area of primary and significant interest will be mailed to you personally.

Complimentary personal subscription
You will personally receive a complimentary subscription to any MCB University Press journal including **Anbar Management Abstracts** in your primary and significant areas of interest if your library takes out a full-price subscription for the lifetime of such subscription. Sample copies available on request.

Regular newsletter
All major developments will be reported and will give guidance on publishing, writing and editing.

Outstanding Authors and Editors
Annual Awards for Excellence will be made to the outstanding contributors among Club members.

Privileged use of Anbar service
For every full-text retrieval required from **Anbar Management Abstracts** a Literati Club members' fee of only £5/US$10 with order.

And more...
We are always open to suggestions of other club benefits that might be added. Please do not hesitate to tell us. Our Authors, Editors and Advisory Board Members are our life-blood. We want to know how we can serve you better. Literati Club is your collective mechanism for feedback.

Table V.
Benefit statement for the Literati Club

offers within the company have been unable to draw on the marketing intelligence clubs now afford and have hitherto been largely imitative of that received through the post for consumer markets rather than in the spirit of industrial business-to-business marketing.

To conclude, it would seem to be the case that MCB University Press has now got three customer-based clubs on Stair 5 of the journey to clubland. It also has a further 70 product/journal subscriber list "clubs" on Stair 1. The scope for movement on that staircase and particularly the intelligent capture of qualified prospects from belowstairs is great. Further developments will proceed on the 80/20 rule of database marketing, however – 80 per cent application and 20 per cent development, or the hypothetical benefits necessary to justify the investment will materialize too late ever to reach the profit and loss account.

References

Beadle, T. (1988), "The care, feeding and deployment of the Lotus User Base", *Industrial Marketing Digest,* Vol. 13 No. 4, pp. 21-32.

Berman, C. (1987), "Cashing in on high tech makes light work of art", *Computing,* 30 April, pp. 18-21.

Blenkhorn, D.L. and Banting, P.M. (1991), "How reverse marketing changes buyer-seller roles", *Industrial Marketing Management,* Vol. 20, pp. 185-91.

Breen, E., Ball, C. and Suffield, M. (1991), "An assessment of the value of introducing an authors and articles database", MCB University Press B-MAP Report, April.

Bruce, B. and Duncan, T. (1990), Hedges, L. and Duncan, T. (1992) (1st revision), *Database Marketing Manual,* MCB University Press.

Capulsky, J.R. and Wolf, M.J. (1990), "Relationship marketing: positioning for the future", *Journal of Business Strategy,* July/August, pp. 16-20.

Chambers, E.R. and Fleischer, R.L. (1986), "How to build and use a database", *Direct Marketing,* October, pp. 128-34.

Cobb, R. (1989), "Beyond courtesy", *Marketing,* 16 February, pp. 32-3.

Cole, T. (1986), "Quality counts above quantity", *Marketing,* 22 May.

Courtheoux, R.J. (1987), "Database modelling: maximizing the benefits", *Direct Marketing,* March, pp. 44-51.

Craig, S.R. (1990), "How to enhance customer connections", *Journal of Business Strategy,* July/August, pp. 22-6.

Curtis, J. (1989), "How an interface database helps Pitney Bowes refine products", *Industrial Marketing Digest,* Vol. 14 No. 3, pp. 33-9.

Ditzler, T. (1987), "The future is database marketing", *Direct Response,* March, pp. 74-8.

Duncan, T. and Hedges, L. (1991), "Database marketing", MCB University Press SEMAP H Report, April.

Durr, J.L. (1989), "The value of the guest register", *Direct Marketing,* September, pp. 50-55.

Giles, W. (1991), "Marketing kinetics", *Marketing Intelligence & Planning,* Vol. 9 No. 5, pp. 4-8.

Grossman, G. (1990), "Continuity marketing – plain and fancy", *Direct Marketing,* March/April, pp. 16-43.

Hotchkiss, H.S. (1987), "Creating a better continuity model", *Direct Marketing,* March, pp. 56-62.

Husani, A. (1987), "Leveraging sales through database maintenance", *Direct Marketing,* March/April, pp. 96-9.

Ives, B. and Mason R.O. (1990), "Can information technology revitalize your customer service?", *Academy of Management Executive,* Vol. 4 No. 4, pp. 52-69.

Johannisson, B. (1987), "Beyond process and structure: social exchange networks", *International Studies in Management and Organization,* Vol. XVII No. 1, pp. 3-23.

Kelsey, R.R. and McGrath, M.J. (1990), "Database marketing targets existing patients", *Health Care Financial Management,* June, pp. 72-5.

Kennedy, I. (1991a), private correspondence within the profession.

Kennedy, I. (1991b), "Tell the media when to stick it", *Australian Marketing,* June, pp. 8-14.

Li, R.D. (1987), "The hunt for direct marketing success", *Direct Marketing,* March, pp. 38-42.

McKenna, R. (1991), "Marketing is everything", *Harvard Business Review,* January/February, pp. 65-70.

Mammarella, J. (1986), "Psyching out list overlays", *Direct Marketing,* February, pp. 46-51.

Michael, P.M. (1984), "Cessna's extensive database zeroes in on flying public", *Direct Marketing,* August, pp. 52-66.

Miles, L. (1991), "An animal instinct for going direct", *Marketing,* 2 May, pp. 25-6.

Nash, T. (1988), "The selling of Kalamazoo", *Director,* September, pp. 105-8.

Ostrager, G. (1986), "Bloomie's database marketing", *Direct Marketing,* October, pp. 94-105.

Passavant, P. (1990), "The strategic database", *Direct Marketing,* May, pp. 40-44.

Pearson, S. (1986), "A lasting relationship", *Marketing,* 22 May, pp. 50-52.

Raphel, M. (1989), "Interchangeable techniques", *Direct Marketing,* March, pp. 42-86.

Raphel, M. (1990), "Take a card", *Direct Marketing,* February, pp. 63-7.

Rapp, S. (1989a), "Here comes the frequent shopper", *Direct Marketing,* January, pp. 70-1.

Rapp, S. (1989b), "RJR vs. Philip Morris: battle of the databases", *Direct Response,* August, p. 26.

Rawstorne, P. (1991), "An entire organization focused on the market", *Financial Times,* 18 April, p. 20.

Roscitt, R. and Parker, I.R. (1988), "Direct marketing to consumers", *Journal of Consumer Marketing,* Vol. 5 No. 1, pp. 5-13.

Rosenfield, J.R. (1988), "The trouble with direct marketing", *Direct Marketing,* October, pp. 40-48.

Savini, G. (1989), "24-karat databases", *Direct Marketing,* March, pp. 36-41.

Senatar, A. (1991), "Selling the arts", *Direct Response,* February, pp. 27-8.

Taylor, J. and Oake, J. (1990), "Maximizing financial services: sophisticated database marketing", *Marketing Intelligence & Planning,* Vol. 3 No. 7, pp. 11-15.

Van Doren, D.C. and Stickney, T.A. (1990), "How to develop a database for sales leads", *Industrial Marketing Management,* Vol. 19, pp. 201-8.

Whitehall, B. (1989), "How Rentokil's affinity club captures professionals", *Industrial Marketing Digest,* Vol. 14 No. 3, pp. 73-82.

Wiersema, F.D. (1987), "Advanced segmentation's practical parameters", *Direct Marketing,* March, pp. 31-7.

Wills, G. (1991), "Enabling customers to drive your enterprise", *European Journal of Marketing,* Vol. 25 No. 4.

Wills, G. (1992), "Enabling managerial growth and ownership succession", *Management Decision,* Vol. 30 No. 1.

Wills, G., Bruce, B. and Duncan, T. (1991), "Creating a marketing intelligentsia", *Marketing Intelligence & Planning,* Vol. 9 No. 4, pp. 1-20 (see Chapter 11).

Chapter 13

Learning in marketing clubland

Database marketing clubs are no different from clubs anywhere else. Existing members need to feel they are getting good benefits; new members need to be recruited. And so it has been for the three publisher clubs MCB University Press (MCB) initiated in 1991 (Wills and Wills, 1992). They serve three distinct segments of MCB's external publics – Librarians in Library Link, human resource management professionals in HR Network and editors and authors in the Literati Club. The very diversity of the three segments makes comparison a powerful approach to improve understanding of what it seems can and cannot be achieved.

Librarians do not normally read the academic and professional journals MCB sells to them. They rely on recommendations from library users, who together constitute a decision making unit (DMU) of some complexity, for advice on how to spend the budget available between the excessive number of requests received.

Human resource management professionals do, on occasion, purchase journals from MCB for the benefit and use of others, e.g. managers on programmes they run, but that is very much the exception. They almost always purchase for their own professional development. Their thirst for outside information from literature is very great indeed for they are frequently called on to advise others on a best course of action. The literature available is a major source of ideas, of reassurance that others have made something work, and through honest case analysis an early warning system of what pitfalls may lie ahead. (All advisory professionals are *prima facie* major personal consumers/ collectors of information in this way.)

Authors and editors are a supply side market for MCB. While there is not a multitude of journals to which contributions can be submitted by academics and graduate professionals, there is healthy competition. To gain the best articles for an MCB journal requires an editor who is both in excellent contact with the leading lights of the field, and efficient/effective at getting these luminaries to write to deadlines. Everything the publisher can do to resource editors in their work and to make the author want to be published in the MCB journal in question must be done if a good perception of quality is to exist and then to prevail. This perception inevitably and very directly links back to the DMU which recommends purchase to the librarians as described earlier.

The authors were partners in the work analysed here but it was, and continues to be, very much a team effort. Colleagues have shared in debate of these ideas at five Comparative Club Evaluation Workshops during the past two years.

With Julian Wills,
Marketing Intelligence & Planning
Vol. 11 No. 11, 1993

As we present and then interpret our data on club recruitment and the benefits provided subsequently, it will be clear that the diversity outlined above has a very direct bearing on what is achievable.

Recruiting club membership

Club membership was defined as gratis and an added subscriber benefit, available to any subscriber/author who returned a questionnaire to MCB giving supplementary information about themselves – information that would enable MCB to enhance the commercial and professional relationship it already had with the subscriber for supplying the high quality academic or professional journal(s) in question. (The club specific benefits offered are discussed in detail later.)

At the outset we had no objective conception of what level of recruitment would be possible. We hoped to do at least as well as Hughes (1992) at Bristol and West Building Society who achieved 30 per cent with mortgagees. However, we bravely hypothesized an ordinal ranking of success that proved to be correct. Librarians would be least likely and editors/authors most likely to respond; human resource management professionals would be somewhere in the middle. We believed this would be the case simply on the basis of what we conceived they could envisage getting out of their club. The authors/editors in many cases depend for their very career progression on publishing, HR professionals for their very career progression on publishing, HR professionals much less so and librarians not at all.

Table I shows the differences we logged, after shoe horning into approximately equivalent categories.

	Library Link		HR Network		Literati Club	
	Number	Percentage response	Number	Percentage response	Number	Percentage response
Total population available	6,776	24	6,288	30	3,500[b]	33
First selection (two or more subscriptions)	2,400	29	585[a]	51	2,500[c]	87
Second selection	4,376	11	193[a]	48	–	–
Third selection	5,324[a]	8	344[a]	11	–	–
Membership recruited at year end	1,625	–	1,865	–	6,000	–
Recommended leads from members	598	37	–	–	Librarians Articles	66 10

Table I.
Recruitment levels of MCB University Press clubs (1992)

Notes: [a]Short form Club questionnaire used; [b]3-year retrospective author file; [c]new authors joining

The first selection recruitment levels, where both Library Link and the HR Network approached existing subscribers who took two or more titles, achieved 29 and 51 per cent respectively. The Library Link second selection was all other librarians, i.e. one or more subscriptions, and netted a much lower 11 per cent HR Network continued its high level at 48 per cent by making its second mailshot to recent new customers (a Welcome Wagon campaign).

There was much early speculation that an important determining variable of lower responses for Library Link could have been the length of questionnaire. It is normally found to affect response. Proceeding to short form for its third mailing, a length similar to that used throughout all mailshots by HR Network, did not significantly affect Library Link's outcome.

Literati Club's recruitment took two forms. The first 3,500 members were informed they had joined! They were retrieved from the three previous years author/editor files; 33 per cent responded to a long form questionnaire about themselves. The second 2,500 members have joined during the 18 months since the club was launched and were required as a condition of having their article printed at all that a journal article record form (JAR Form) be submitted with the article. Eighty-seven per cent was achieved and the shortfall was largely attributable to editorial neglect and rushed publication schedules.

One final comment on Table I must be made on lead generation, Library Link specifically asked for the names of individuals in the DMU affecting MCB journals, and received responses from 37 per cent of the 33 per cent who returned the questionnaire. Literati Club asked for the name of the author's librarian and this was supplied in full by 66 per cent, with good leads for further articles from self or colleagues being provided on 28 per cent of questionnaires.

Next steps in recruitment
It will only be readily apparent that the cell sizes within each club are very small by normal standards. The direct incremental average costs of recruitment for each have been surprisingly similar at £10/£20 per member. The average cost continues to fall as numbers rise since, apart from testing new recruitment letters, the major expenses of design and learning do not recur.

The next steps for Library Link and HR Network, during 1993, are to seek to build much higher levels of recruitment both by further incentivization, and by strategies similar to that adopted in Literati Club. Subscribers will automatically become members of the appropriate club. But they will be given differing levels of service depending on the extent to which information is provided to permit focused promotional campaigns, that will increase and/or retain the income derived from the member in question. The assumption we are making which has proved valid thus far with Literati Club is that those who, for whatever reason, did not find the time to complete and return the informational questionnaire, either in long or short form, are nevertheless as likely to be interested in the benefits of Library Link or the HR Network as those who did. Only time and testing will tell whether they are more or less so.

What the clubs offered

It will be recalled that recruitment took place only among existing subscribers to MCB journals/existing authors. As such all were already familiar with some products and services of MCB – although none would be familiar with anything approaching a majority even in their own speciality field.

The major perceived benefit of Library Link membership as events turned out, was the second item listed – prepublication details. Every month in the areas which librarians had pinpointed as of significant interest to them, all the appropriate advanced contents pages, abstracts and keywords were mailed in a handsome folder to members. The service incurred a direct incremental expense of £21 per member per annum. All other benefits in the offer were in comparison very inexpensive indeed.

HR Network provided a similar advanced contents pages, abstracts and keywords service for its members, while at the same time promising free trial copies of journals, not specifically requested, in identified areas of interest within HR. Furthermore, the HR Network was launched gratis for charter members but with a £50 p.a. membership fee envisaged for later entrants and for any who cancelled their MCB subscriptions but still wanted to belong.

In both cases, therefore, the major substantive offer was a free Current Awareness Service, albeit the launch promotional copy gave no differentiation or prioritization of one benefit against any other – as we now see with the benefit of hindsight. This same lack of differentiation or prioritization applied to the offer to Literati Club members. Judgementally, the major benefit to its members would seem to be the readily open access for getting potential contributions considered, followed by the complimentary copy of a journal "if your library took out a new subscription to it and if you helped to get that MCB journal on the shelves".

One mild anxiety that Literati Club had at the outset was whether its name did not overgild the lily a trifle. But it proved not to be so. Quite the reverse. Authors and editors were amused and delighted to be seen as Literati, and the Annual Awards for Outstanding authors and editors symbolized MCB's respect for people it seldom meets in the course of its academic and professional publishing. Whereas Library Link and HR Network were, in a sense, just one more commercial/professional grouping, Literati Club was a unique notion that created a sense of identity which clearly made authors and editors feel good about their publisher, and its concern for their wellbeing on the supply side.

The additional intelligence sought

As was observed at the outset of this article, clubs prosper if they serve their members well. To achieve good service, good intelligence is required. So the strategy of creating clubs was to enhance what was already known, i.e. MCB knew the names and addresses of the individuals targeted, and either the journal(s) subscribed to, or for which an article had been written – or at least it thought it did. Table II indicates the astonishing extent to which basic updating was required in Literati Club. Comparable data was not analysed for the other two clubs but was estimated to be in a less parlous state by far because of the regular renewal processes.

	Percentage
Address correct to club member's satisfaction	39
Marginally incorrect, i.e. job change, spelling, postcode omitted	31
Substantially incorrect, i.e. wrong postcode, address or deceased	30
Phone or fax numbers provided by members as requested *de nouveau*	86

Table II.
Getting the demographics
right for Literati Club
at the outset

All club joining questionnaires were designed primarily as vehicles to collect additional intelligence, all sought confirmation of correctness of name and address details and job titles, then went on to collect *de nouveau* telephone and fax numbers. Beyond this point, HR Network took the short form questionnaire route, with a cash incentive for speedy return. The short form solely sought out intelligence on the subscribers' areas of interest within the sub-fields of HR, e.g. career management, organizational change, training, managerial psychology. This additional information was of direct potential value because MCB had niched titles in most of the areas – and if it did not, scope for new launches could well manifest themselves. As we have seen in Table I, response levels were most considerably higher than the long form questionnaires sent to librarians when comparison was taken between those with two or more MCB subscriptions already.

Library Link's questionnaires gathered in a necessarily much wider field of interests from members. Librarians, after all, were normally responsible for the entire stock of management journals across all disciplines and professions. No limit was placed on how many fields a librarian could indicate as significant. But considerable further detail was elicited on the DMU within the library concerned, on its budgetary cycle, recent trends therein, the overall stocks of journals and the total annual spending levels. The short form questionnaire, adopted for the third selection, being a second recruitment drive to the total population, did not, as Table I showed, dramatically transform uptake. The loss of potential intelligence for subsequent purposes was, however, considerable.

Literati Club intelligence gathering was the most extensive of all. The spread of authors' literary endeavours with MCB and elsewhere was measured, so too were areas of intellectual interest, use of Anbar services, images of MCB as a supplier, willingness to submit future articles on disk etc. The very high levels achieved ensured extremely good understanding through the long form questionnaire used.

Customer development with the benefits of clubland intelligence
Human Resource Network
The HR Network's strategy of short form questionnaires meant it learned the least additional intelligence beyond what MCB already knew. Nonetheless they were able to focus wonderfully on putting it to good use. Table III identifies the main interest of club members as recorded in the questionnaires. This information meant they had identified for a sizeable proportion of higher value customers what their complementary fields of interest were. Promotions were accordingly carried out on the basis of this intelligence throughout 1992 and continue as more customers become club members and/or new journal titles are launched/acquired.

New launch thinking is naturally influenced also by this intelligence, bearing in mind that your most likely new customers are those you already have. The new *Journal of Recruitment Selection & Retention* was, for instance targeted to club members who indicated these areas as of major interest.

The advanced contents pages, abstracts and keywords service, which was the most visible basis .or regular contact with club members led to requests for journals from 320 members (844 samples) during the first 12 months, leading to

	Percentage
Management development	10.6
Leadership and organization development	10.1
Organizational change	9.48
Education and training	7.73
Performance appraisal	7.18
Managerial psychology	6.91
Career management	6.14
Skills training	5.00
Employee counselling	4.42
Manpower planning	4.36
Women in management	4.30
Vocational training	4.17
Recruitment	3.74
Retention	3.04
Employee relations	3.04
Pay remuneration	2.27
Health and safety	2.27
Personnel administration	2.03
Industrial relations	1.81
Others	1.38

Table III.
Main interest areas of
HR Network members

4.7 per cent (1.8 per cent) new sales per customer (sample). Although this clearly more than covers the expense of sample copy mailings it was not at all certain that the major expense of mailing the advanced contents pages each month could be justified overall. A quite different major visible feature should perhaps be sought. One was later tried – the launch of a new journal, *Human Resource Management International Digest,* targeted directly to the HR Network members as their journal at a discount. It was also offered openly with the same discount available if new subscribers joined the HR Network by returning the necessary questionnaire. In total, 160 (9 per cent) took it up – a most considerable advance on sample copy results by April 1993.

This notion of making club membership into an offer to new subscribers was nowhere more rigorously pursued than in the HR Network. When used with 1992 new subscribers to MCB's titles under the house banner "Welcome Wagon", it created additional sales to 6 per cent of them – vindicating the "recency" dimension in direct marketing.

The combined outcome of requested sample copy despatches, the mailing of unsolicited samples in interest areas in the same folder as advanced content pages and the Welcome Wagon/club journal approaches, was that *10 per cent of existing customers acquired a further journal subscription during the first full 12 months.* The total cost of gaining them, including all club launch expenses, was well covered by the lifetime value of the subscriptions gained. However, there was no comparative data for the levels of response to non-club promotions previously – important data that MCB has now set its hand to discovering from the fortuitous arrival by acquisition of a suite of north American marketing titles. Rather than forming an "official marketers club", intelligence from each subscriber will be gathered in on all new order forms and at renewal time.

This comparative "unofficial club" initiative managed in parallel to the HR Network team, is of vital future significance. However, since the HR Network has obviously been worthwhile, there has been no faltering in plans for its further development. Gold and standard levels of membership, for those who have provided the intelligence needed, continue through 1993 to receive a very full service. Those who have not (to be known as basic level members), but who are MCB subscribers in the human resource area, will perceive the image of the HR Network but receive no greater benefits that the "unofficial marketers club" described above.

A 1993 target level for new sales to existing customers of 7 per cent rather than the 10 per cent achieved in 1992 has been set for all 5,895 members at all levels of service from gold to basic. A budget of £100,000 is being deployed in search of gross subscription income well in excess thereof.

Library link
Library Link's intelligence gathering, it will be recalled, was far more extensive than for the HR Network. Nonetheless it had lower proportional recruitment success. Librarians, however, have a much wider range of interest areas than human resource professional managers, and much greater scope for focused

promotional effort was available to each member. Table IV identifies the most frequently chosen interest areas of librarians joining Library Link.

Library Link's advanced contents pages, abstracts and keywords service generated requests during 1992 for sample copies from 527 members (1,999 samples). These led to a 2.8 per cent (0.7 per cent) new sales per member (sample).

This was a source of even greater disappointment, than with the HR Network where the levels were 4.7 per cent (1.8 per cent) respectively. (It included additionally a Publication Date Information Service for librarians, telling which issues of MCB's journals had been published and how many were still due.) Despite the obvious and oft stated appeal of the total service to librarians, there was little likelihood that it was the most cost effective promotional approach available.

However, Library Link had distinctly better results than the HR Network on its promotional activities to areas of significant interest as gathered from the questionnaires returned. Because of the ordering processes used by most librarians, i.e. agents/booksellers, it proved impossible to trace the precise source of all the new subscriptions gained. However, *the share overall of Library Link members taking out a new subscription came near to twice the level of the HR Network, at 18.2 per cent.* And because of the higher average coverage price of the typical library purchase, the gross revenue was some five times higher – achieved with a not dissimilar level of investment.

It must be reiterated, yet again, that librarians have a very broad based constituency of readers and much larger budgets than the typical human resources professional manager. MCB has some 100 titles to offer the librarian that they do not currently take and in every case several within the specific focus of significant interest.

It has also been suggested that Library Link members received other promotional impacts during the year recommendations from users for instance. And they surely must. But the differential result is so great it calls for determined strategies for the future.

	Percentage
Information management	87
Librarianship	86
Information science	81
Informations systems management	79
General management	78
Personnel management	76
Training management	74
Management development	72

Table IV.
Most frequently chosen
areas of interest for
Library Link members

The primary goals for 1993 have been set to increase Library Link membership in the same manner as HR Network, i.e. by offering a basic level of Library Link membership for the entire population of librarians currently nearing 7,500. The secondary goal is to bring under control the escalating expense of providing the highly visible, much appreciated, but seemingly cost ineffective advanced contents pages, abstracts and keywords service.

To meet the challenge in this second respect, it has been resolved to provide a Library Link Electronic Bulletin Board Service in place of the monthly mailings from August 1993. It contains the cost but, more importantly, creates considerable opportunity to integrate the Library Link relationship with its members into the embryonic range of MCB electronic publishing and Anbar Management abstracts/full text library services.

This latter, lateral thinking has been a hallmark of all club activities since their inception in 1991. It is a point to be returned to later in discussing the strategic significance of the close encounters with customers that database marketing clubs inevitably ensure.

At the outset, it was mentioned that extensive intelligence was gathered from librarians in the long form questionnaire. One element was the names of other influentials in determining library purchases – and targeted mailings were made to these with good effect. Information was also collected on the decision cycle for new subscription ordering by libraries. The myth that purchases are made once each year was exploded – except in Europe, the great majority of decisions were being made around the year. The outcome of this was included in the tactics of timing promotional campaigns everywhere.

A Library Link newsletter, not a journal, was also introduced as it had been for HR Network. The newsletter helped to develop the overall image of MCB as being directly concerned for the work of the librarian – and through Library Link and its freephone number, just a phone call away to a named individual. Such soft benefits are impossible to quantify, and are the justification the world over for building overhead expenses. The search was accordingly on for some hard tangible results in renewal levels. Some early indications were forthcoming when Library Link membership was used as a telechaser incentive in the renewal cycle of both floppy Anbar and full hard copy text subscriptions.

Literati Club
The establishment of the Literati Club within MCB since 1991 has been one of the more astonishing of all recent events. The idea was conceived by sub-editors originally in the production department as a more orderly way to keep demographic data on authors and editors, for chasing late copy, and despatching proofs and finished journals. Truth to tell, MCB had never sought to manage the area on anything approaching a professional basis – several individuals held their own shoe-box records and the full list on the customer services database was found to have the startling omission rate of nearly half of all Editorial Advisory Board members! So it was not surprising that once the

issue was raised a tidal wave of ideas washed over it. It was indeed hijacked out of production department into a separate unit off site.

We have already indicated in Table II how poor the demographics were. But we were able to pinpoint speedily who were MCB's most prolific authors, in what field they had written by keyword and journal title, including recency of their contribution.

Fully 28 per cent of club members report a new contribution is under preparation which they want to be considered in due course by an MCB editor – over 300 had been passed on by the end of 1992.

The authors and editors were asked to identify their areas of professional interest over and beyond the keyworded contributions in the database. Calls for Papers in specialist fields can now be speedily targeted on a broad scale all over the world.

Of particular significance, authors were asked to detail the name of their "favourite" librarian for cross selling both the journal in which the author's article had just appeared with a totally focused letter and sample copy and, after de-duplication, Anbar management abstracts. Two-thirds of authors willingly gave the librarian detail needed.

The final batch of questions gave a clear direction to MCB on how to improve its services to authors as set down in Table V.

Less than half the authors proffered any advice at all – they were satisfied or unconcerned – but multiple responses flowed from the others. The Literati Newsline carried full details of the findings and editors were alerted to the ways authors were looking for help. Editors' workshops were convened and a briefing document "as the Editor said to the Author" created. Over half of all authors tick the box asking for their personal copy. Considerably more can and will be done (Breen, 1992).

One particular ghost was laid by the questionnaire – the willingness or otherwise of authors to oblige their publisher by providing their articles in an electronic format – either by sending a disk or scanning. The majority of authors in all parts of the world agreed it was perfectly reasonable for a publisher so to request – north Americans agreeing 74 per cent to a low of 50 per cent in the UK.

By the end of 1992, over 100 new journal ideas had been forwarded from club members to MCB's business development unit for consideration. Ninety per cent of librarians named by authors were targeted for Library Link membership, and fully 80 per cent as well qualified prospects for journal promotions in the field of the author's work. The fourth Awards of Excellence for authors and editors – the first under Literati Club aegis, had been held in London with an accompanying well attended editors' workshop. Finally, over 100 editors and Editorial Advisory Board members had agreed to assist MCB to promote the sales of its journals at conferences and workshops they attended in the course of their normal professional work.

In post-publication surveys, 98 per cent of authors published in MCB journals rated the overall level of service received as satisfactory – only 2 per

Recommendations	Total	North America	Percentage ROW	Europe	UK
Speed up review and publication	8	14	10	10	5
State clearly and enforce review procedures	6	7	5	2	5
Reconsider pricing policies to expand readership	6	9	5	6	3
Publish formal editorial plans	4	3	4	4	5
Circulate portfolio of MCB journals	3	5	4	5	1
Convene author meets editor meetings	3	2	1	4	3
Provide gratis off-prints	3	3	4	0	2
Pay the authors	2	1	2	1	3
Improve quality of author proofs	3	2	1	6	2
Offer accelerated production of research notes	2	1	2	4	1
Include more practitioner articles	1	1	1	1	1
Miscellaneous	12	17	18	16	8

Multiple responses were permitted No 1,377

cent reported it could have been better! Hard to believe. The major constructive criticisms offered returned again to the delays in publication arising from the review system and the occasional editorial hold up.

Literati Club has just about everything going for it. Authors want to be published; editors want as much assistance as they can get; MCB wants good quality contributions in on time. So the Literati Club with virtually no hard benefits quantified to date – either from the Library Link promotions or the editor's willingness to assist sales, has developed a gigantic halo effect. It is loved among all seeking to do their administrative work more effectively, and there can be no doubt that soft productivity improvements abound. An unseemly withdrawal in 1992 by a disgruntled editor and his entire Advisory Board, to launch a competitive journal could have been a show stopper. But Literati Club probably paid for its entire costs in that single moment of truth. Authors were assembled, and a new Advisory Board, and to cap it all a new launch straight into the defector's niche market. It has so many lovers it is hard to know which the club should respond to next. There is extrapolation to be done of existing tasks and there are committed tasks to do better. But the real potential has only just dawned in the minds of senior management

Total marketing reconstruction
It is frequently claimed that for any major initiative of any sort to succeed in any enterprise it needs top level support. Well yes. Even though, as described

elsewhere (Wills *et al.,* 1991), the origins of MCB's database marketing were bottom up albeit mentored from top down, on this occasion learning from clubland has been very much a case of top down. The conceptual message from clubs at large and from the wider "unofficial club" workings of MCB's total marketing database, is that MCB can with very great benefit, reconstruct its marketing organization, and put in place major teams to exploit the potential for developing sales with existing customers. Fletcher *et al.* (1992) argue appropriately aligned structures are the only way to gain the major benefits. Jackson (1991) identifies from the Blue Cross just how difficult they are to achieve. The reconstruction is now described in MCB as "Customer Development", and from the 1 January 1994 will be operational in two divisions – for librarians (22 per cent of customers) and for the rest of our customers (78 per cent).

A second major conceptual awakening, top down, has already been hinted at. It arose as the Literati Club got into its stride. It became crystal clear that a major source of "quality" in MCB's journals was to be found in the logistics flow of articles from the authors through editors into production and outwards to librarians and readers. Whether the whole or part of that journey be made electronically or as hard copy is obviously important, but cannot outweigh, or be allowed to distract from, the omnipresent need to build on and develop the supply side flow of contributions into MCB and the screening and selection processes thereof.

The received organizational wisdom of MCB from 1977 to 1992 was that journals were clustered around a few publishers who were wholly concerned with the editorial inflows, the promotional campaigns to existing and potential customers and the profitability of each journal. From 1994 onwards the organizational vision is of three customer focused business units:

(1) the publishing logistics flow from author to reader;

(2) the servicing and development of existing customers; and

(3) the prospecting search for new customers not yet in the MCB marketing database.

To accomplish such a total marketing reconstruction at MCB, all senior and middle managers are again engaged (Wills, 1992) on at least nine months of action learning studies to see and plan how the vision can be implemented.

The centrepiece for customer development is an integrated customer database – serving the needs of existing and incoming customers, financial services and the distribution centre to go live in March 1994 on the far side of 1994 volume renewals. Prospecting too has a keen interest because the new database will facilitate profiling of potential customers as well as be a more accomplished customer services interface with new customers as found. It will also provide the basis for MCB's first automated campaigns to its customers, most immediately for the Welcome Wagon which the club has already demonstrated addresses one of the most potent opportunities both for additional sales and for intelligence gathering over and beyond product data and demographics. The benefits are expected to be well in line with IT norms of 10 to 30 per cent sales improvement (Moriarty and Swartz, 1989; Morris-Lee and Erickson, 1991).

The specification and design team is drawn from all the interested parties involved. They are interfacing directly with the software suppliers aided by MCB's senior information systems staff. Such a high level of participative management is intended to ensure the highest possible levels of ownership by March 1994 as (widely) recommended (Gib and Margillies, 1991; Luthans, 1991).

The centrepiece for prospecting activities within MCB is to be regionalism. For the first time, it is almost too surprising to tell, the potential sales among new customers will be assessed on a country-by-country basis – not journal-by-journal or list-by-list. The concept was pioneered in Australia when it began to look at Far East market opportunities in the early 1990s. It has enabled the Brisbane-based Sales Region to open up new markets on a carefully matched basis from the total MCB portfolio of journals. Returns from the promotional $ spent in the Far East on this basis are more than twice as good as in Australia – aided it must be said by the economic growth of the Asian Tiger countries in contrast to recession hit Australia. Full time sales offices in Japan, Singapore and Brisbane, and agents in Hong Kong and Malaysia provide a model to be examined for the rest of the world. The psychological dominance of a Head Office in Bradford that, inevitably, consistently sees the "home UK market" as the best first move for any sales activity, is to be challenged by the spirited advocacy of Regional Sales Development Managers for the rest of Europe, for Africa, India, north America, the Far East, Australia, Japan and South America, each reporting to a Board level executive.

The centrepiece for publishing logistics is a marriage between two parties already half way down the aisle but only now being formally introduced to one another! The two parties are the technological process that takes authors words and transforms them into electronic and/or hard copy outputs, and the Literati Club's delivery of a focused strategy for relationships with authors and their editors.

Emerging strategic perceptions of clubland
We have thus far discussed in detail the creation and cost beneficial growth and development of MCB's three clubs. We have pinpointed how their presence and messages were catalysts to a reconstruction of the marketing organization of MCB. It is appropriate now to reflect on our emerging strategic perceptions of clubs *per se*. Ward (1987) argues they are inevitable.

The greatest merit of all in a club is that it brands not a journal or even a group of journals. It brands a group of customers. As such, marketing clubs can be a powerful tool in the armoury of customer orientation. Furthermore, if clubs have sufficient critical mass they induce exactly the interactive approach with their members that is at the heart of any successful enterprise.

The great danger of clubs is that they become fascinating in their own right for managers in the enterprise, satisfying a desire for information, with little attempt to assess the hard, or even value the soft, benefits to be traded against the costs incurred. Customer interactive benefits are replaced or swamped by risk aversion and reassurance behaviours.

Visible, branded clubs should accordingly never be any more highly valued than the moving average response they gain to promotional efforts. It does not have to be a financial response necessarily – although finance is a wonderful deceiver. But there must be an objective function that is routinely monitored and publicized that keeps it all in a sensible commercial perspective.

If such perceptions are valid, it implies that as the visible club grows and develops, the utmost creative attention must be given to the quality of the branding and the identification of the objective function(s) to be optimized.

Branding clubs
As the detail analysed earlier shows, the branding for all three clubs was intuitive/judgemental/experiential – definitely not objectively arrived at:

- Human resource professional managers were assumed to like the notion of being members of "The Worldwide Literary Forum for HR Professionals" called Human Resource Network.

- Librarians were assumed to like the notion of "Helping Us to Help You", and join Library Link.

- Authors and editors were assumed to like the notion of compulsory membership of the Literati Club, helping "Editors to help Authors help Editors".

MCB's intuition and experience are obviously not far wrong – but how much more right could they have been, and how much more convincing creatively could the mailers and the folders and the Newsletters have been, if the whole endeavour had been perceived as branding?

Further, how much more effective could the club benefits have been/can they be if we elicit from members what would be most useful?

MCB's benefits were products and services it was convenient to offer. We had to start somewhere, but where next? What approach can impact most on the objective functions we are seeking to optimize?

The objective functions of subscriber clubs
What truly should be the objective(s) of each subscriber club? They certainly cannot all be the same, albeit that they ultimately, over a time scale, must be cost effective?

Potentially, HR Network, and Library Link went *ab initio* for the hard-nosed bottom line. How many new subscriptions could be cross sold to existing customers? As we have seen the answers were good to very good indeed. It could perhaps have seemed a high risk strategy at the outset, but with hindsight it was no real risk at all. The major benefit was it introduced a simple objective function up front, and was dismissive of soft, friendly goals.

But it was the best objective function to address? And if so, what other objective functions are also suitable to be worked on as well? The most obvious is, of course, subscription renewals. Could clubs not have been targeted to improve the level of their renewals? Human resource professional managers have an average renewal level that is at least 15 per cent inferior to that of librarians for example. Would a similar level of dedicated activity have given better financial returns than the new subscription sale as the focus? We can only find out by doing it. The explanation for the choice of new sales, however, was a relative sense of satisfaction that MCB's Renewals Improvement Campaigns (RICs) created as a result of the first senior managers' action learning programme in 1988. With some 2,000 human resource professional subscribers being lost each year through non-renewal even after the RICs campaigns, there must be good pause for thought. Far more pause for though indeed than for librarians where not only are cancellations fewer – a loss of only 1,250 each year – but the basis for the cancellation likely to arise from a DMU and to be almost certainly irreversible in the short term.

In fairness to HR Network colleagues, they did indeed turn attention to renewals in the final stages of RICs. But it was not a high order priority and no targets were set to be monitored – just an optimistic belief it should help. (An ominous sign of the dangers of clubs as already suggested.)

A third objective function which has emerged during the clubs first two years has been the scope for "Key Customer Account Management", including "Customer Account Profitability Analysis". No customers currently receive any discounts for quantities purchased of the same or different titles other than as an initial incentive to subscribe to a title. Most particularly among librarians, but also surely among human resource professionals in major enterprises, scope must exist for negotiated quantity arrangements, that seek to manage not the contribution or profit levels per title or product line but per customer. Some librarians, for instance, take up to 21 titles.

Every serious Graduate School of Management in the world is a potential customer for a broad portfolio of MCB titles. No effort whatever has been directed thus far at seeking to develop major customers – actual or potential. There would seem to be ample justification for both in customer development and in prospecting.

MCB's branded subscriber clubs should look very closely at all their objective functions to identify what balance of activity offers the best medium-/long-term potential. Analysis thus far shows that clubs among direct readers, such as HR Network have less scope for selling new subscriptions but perhaps greater scope for renewals improvements than Library Link. Library Link offers considerably greater scope for selling new subscriptions because of its members much wider constituencies and budget provisions.

Both clubs, but most particularly Library Link, have scope for Key Customer Account Management.

The "dilution of club membership" to include all subscribers whether or not full information is available may or may not succeed in eventually bringing more intelligence at the levels required into the club databases, i.e. the move from Basic to Standard to Gold for HR Network and from Basic to Standard for Library Link. There seems little danger in the dilution and every hope that the branding available already will enhance responses not only to requests for information but new subscription sales as a result as well. Unless this objective function, or renewals, or key customer account management are more than proportionately improved, (with monitoring formally put in place to measure it) then dilution will have been misplaced activity. Such customers could equally well be left as an "unofficial club of those reluctant to provide additional information".

Finally, the phenomenon of the unofficial club must be thoroughly tested and evaluated. Without visible branding it is in grave danger of being an occasional rather than a sustained activity, with little hope of addressing any objective function other than new subscription sales.

Objective functions of Literati Club
Despite spirited attempts in Literati Club to make new subscription sales an objective function, no dedicated infrastructure was in place to do so.

Library Link picked up the leads for authors' favourite librarians but no data has emerged on its effectiveness thus far. The drive to get editors to actively promote sales for their own journals at major professional events was resisted within the prevailing organization structure for half the time under evaluation. Again no data has since emerged and no campaign of follow through has been conducted.

The soft objective functions won the day hands down. Head Office business developers, and executive publishers short of copy, queue up to make use of the database for extremely worthy goals. Some 30 calls were made in the first full year and that is only the beginning.

So convenient and helpful has Literati Club been as currently constituted that its far greater future potential was at one time in danger of being overlooked. That greater future lies squarely in the field of improved content

	HR Network	Library Link
Club population at year end	1,865	1,625
Advanced contents/abstracts/keywords service		
Members asking for samples	320	527
Sample copies requested	(844)	(1,999)
New subscriptions gained		
Per member	4.7 per cent	2.8 per cent
Per sample	(1.8 per cent)	(0.7 per cent)
Welcome Wagon new subscriptions gained	6 per cent	n/a
Total new subscriptions gained during year with club members	10 per cent	18.2 per cent

Table VI.
Customer development achievements in 1992

quality in each issue of a journal – both per article and from the selection offered. By any objective analysis 28 per cent of all journals/editors suffered from shortage of what they perceived as good copy for their next issue, i.e. at copy deadline they had too few contributions available. A further (by deliberately solicited in-house views) 25/50 per cent of titles obviously have scope for improvement, not being what might be described as the leading journals in their field. Of the balance, only 10/15 per cent of titles could honestly be described as seldom, if every, having a poor article or selection.

But there is no objective evidence available of what journal authors or even their editors think of the quality of MCB titles beyond an occasional, less than scientific published survey from a professional grouping or a doctoral student. Only at the time of cancellation do we learn at MCB that one of the major reasons for cancellation is that the content was "not what the reader had expected".

Literati Club is accordingly strategically placed for MCB to optimize at least two objectives that have not so far been addressed.

First, it can greatly improve the efficiency and effectiveness of editors at attracting and gathering in articles and selections by increasing the flow of articles. This enables choices to be made, rather than the current reliance on a good editor doing a good job without a great deal of help from MCB, unless he is obviously out of his depth. MCB has for a long while provided written guidance to editors entitled "As the publisher said to the editor", but there is no measured evidence that it has improved anyone's effectiveness. Focused assistance at appointment, then later after induction for the poor performers, with eventual changes to get matters right is required and can be accomplished.

Second, Literati Club is also strategically well placed to extend the Awards of Excellence each year to be a continuous monitoring of excellence/quality during the year. The criteria that editors and Editorial Advisory Board members use to judge a best article of the year will surely differ from the concept of the excellent/quality journal – but the club's population of authors and Editorial Advisory Board members can be invited to identify the criteria, and to make the judgements. These are excellent roles for Advisory Boards to play (Mueller, 1988). Even though it must be accepted that only very poor response levels would be achieved among readers to queries on journal excellence/quality, there seems every reason to believe authors will also respond as well to requests for intelligence on this issue as they have done previously for Literati Club.

Finally, Literati Club has the same dilemma as HR Network and Library Link in terms of wear out among existing members. Introductions of new authors, seeking out new editors and new Advisory Board members should also be evaluated as an objective function to be optimized. It is something where all three clubs can come together in common cause – not in a cold prospecting format but on the basis of "introduce a colleague". Thus far, Library Link has led in this respect with some 598 leads to the end of 1992.

Back to basics

The surprising course of this paper thus far, from basic issues on setting up marketing clubs and measuring their initial success, to reconstruction of the organization itself, and on to revised strategic perceptions of clubs, suggests we should return to basics to conclude (Barker, 1993). At the operating levels within clubs, what key areas of professional improvement are needed? How can MCB's proprietary persuasion processes (Roscitt and Parket, 1988) be burnished?

(1) *Electronic interaction* – both Library Link and HR Network have been in danger of allocating the great majority of their promotional budgets to photocopying and postage costs on advanced contents, abstracts and keywords – which have been demonstrably under achieving on the objective function of new subscriptions sold. As such 1992/1993 has seen the indepth exploration of Bulletin Board Systems to replace the monthly despatches. It raises a wide variety of issues of how to get the best interaction from customers, and how to use complementary paper, telephone and fax interactions to ensure the Bulletin Board is not a white elephant. However, there is widespread evidence of it at work successfully elsewhere with both librarians and human resource professionals.

Such thinking also has very wide scope among Literati Club editors. The monitoring of copy status for an entire volume in advance by MCB will enable it to identify where quality problems might happen before too long – and enable action to be taken before the problem meets a publishing deadline. It also offers great scope for the interaction of editors and executive publishers at MCB on proofing, selections of articles in each issue, and the implementation of the annual quality assessment cycle/Awards of Excellence.

Several editors already track their refereeing procedures on their own initiative with built in time checks. MCB can interact with these procedures, offering the appropriate software to all with guidance on editorial procedures via a HELP function (Gimson, 1993).

(2) *Data integrity* continues as a perennial challenge most particularly at the customer service interface with customers and the prospecting interface. It can be turned to good sales advantage with appropriate training (Robbins, 1991). These are key moments of truth, just like Literati Club's forthcoming publication date for an author, when collaboration and assistance from the customer are at a maximum. The organization dilemma is always that the benefits almost all flow to others elsewhere in the enterprise rather than to the individual asked to ensure the integrity. After five years endeavour, MCB is still unaware of 32 per cent of its customers' job titles and 14 per cent of their industrial classifications.

(3) *Creativity and copy writing* are, also, a perennial challenge. An in-house Best Campaign Award system was launched in 1992, and the Australian and Far East offices have introduced a monthly office hot

shop/peer group best campaign of the month review. The challenges, are to identify not only what to say that matches the interests of the potential reader or club member, but how to say it powerfully and well. With upwards of 2,250 campaigns conducted in 1992, all on a relatively small scale, it is a giant challenge. Training workshops are the order of the day because testing on such small cells leaves good field measures often impossible to achieve. Bold and imaginative ideas are need (Spedding, 1988).

(4) *Profiles analysis* of existing customers is long overdue to identify renewal danger areas for special encouragement to stay, and to identify differing propensities of new purchasers. The reassembly of MCB's marketing data in the Integrated Customer Database by March 1994 must trigger a sustained analytical approach to the customer data we have. Good news like the "Welcome Wagon" works well, and "Gotcha" for new subscribers from acquired titles, should be developed in less likely but more focused ways with good statistical analyses. "We Want You Back" campaigns (Eisenhart, 1992) for lost customers, should be targeted and incentivized in an appropriate tailored fashion.

Profiles will also form a vital clue for the development of prospecting activities (Schnell, 1992). Plans for a Library Link "Shadow club" of qualified prospects by profile have already been made and will advance under the new marketing reconstruction as a top priority. Rather than offering journals by title, the major qualified prospects will be approached as a potential Key Customer Account, with "full service arrangements" proposed. Some 5,000 librarians worldwide are sought.

Finally, what are the characteristics of samplers for MCB journals? What types of sample takers end up as subscribers and which not? Can we ask for more intelligence before we send a sample copy out – as our prerequisite so to speak – that will help us learn?

(5) *Frequency of impact* requires most thorough evaluation. The wear out or breakdown effect is frequently inadequately understood if at all, and prejudiced opinions prevail at total variance from what data exists. Copy can be varied to the same list with great effect, thereby permitting higher frequencies (Berry, 1993).

(6) *Private database media* such as the club newsletters need careful design and evaluation (Cross and Smith, 1992) to meet the necessary image building and communication goals that must be deliberately set. Even the most lavish can falter on cost effectiveness criteria within a total portfolio (Royal Mail Streamline, 1993).

(7) *Renewal communications* require creative evaluation as well as the continuing review in RICs of mechanism and offers. Brock (1988) argues they should unashamedly sell hard. In the mechanism field, however, some fascinating new initiatives are occurring among subscription

agents (Basch, 1988), for electronic data interchange, and outside the agent framework. Australia Post for instance (1993) has introduced EDI Post for all state capitals which includes cash collections.

Action focus points

It is relatively simple to conclude that marketing databases have the potential in any enterprise to be a learning dynamo. Their interactivity with customers when the data is suitably analysed creates a never ending flow of ideas and opportunities. The managerial challenge is how to control or lead the enterprise forward in a cost effective manner (Anderson and Hoyer, 1992). It has been argued that the ferment arising from customer interactivity must be translated into a finite number of objective functions to be optimized at any one time. Then management, and indeed systems development, can focus and learn as opposed to being bemused.

MCB has practiced this self discipline every since it introduced its marketing database in 1989 (Wills *et al.*, 1991) and will continue it. Continuous innovation and improvement is not the order of the day. Step function improvement is.

Once the self discipline has identified what can be appetisingly described as munchable chunks, action teams and plans can be well briefed and performances against objective functions measured. Goodstadt and Marti (1990) were adamant on this point in their work with National Westminster Bank focusing on its transformation of customer service.

The self-disciplined action focus points for MCB in the next 24 months are to:

(1) see the reconstructured organization into place to enable strategic intent to be accomplished;

(2) see the centrepiece systems for each new business unit crafted and operationalized through the mechanism of action learning projects for senior managers;

(3) sustain business performance during the transition phase to reward which profit-related pay has been instituted;

(4) enhance the professionalism of the operational staff for which once again action learning programmes and projects have been instituted from April 1992 to October 1993 for junior managers and supervisors.

Notes and references

Anderson, W.T. and Hoyer, W.D. (1992), "Marketing in the age of intelligence: the case for control", *European Journal of Marketing*, Vol. 25 No. 8, pp. 32-54.

Australia Post (1993), "Cutting time and distance by mail", *The Australian*, 12 April.

Barker, J. (1993), "Developing and using database marketing for an international publishing house", *IT for Marketing*, UNICOM, Brunel University, 19 March.

Basch, N.B. (1988), "Checking out the library market", *Circulation Management*, May, pp. 30-3.

Berry, J. (1993), "Introducing database marketing at the Imperial Cancer Research Fund", *IT for Marketing*, UNICOM, Brunel University, 19 March.

Breen, E. (1992), "Journal publishing: how editors and authors work more effectively together", *Marketing Intelligence & Planning,* Vol. 10 No. 11, pp. 38-40.

Brock, L. (1988), "Renewal letters should sell hard", *Direct Marketing,* June, pp. 80-2.

Cross, R. and Smith, J. (1992), "Staying in touch with private media ", *Direct Marketing,* June, pp. 28- 32.

Eisenhart, R. (1992), "We want you back", *Business Marketing,* August, pp. 37-9.

Fletcher, K, Wheeler, C. and Wright, J. (1992), "Success in database marketing: some crucial factors", *Marketing Intelligence & Planning,* Vol. 10 No. 6, pp. 18-23.

Gib, A.G. and Margulies, R.A. (1991), "Marketing competitive intelligence networks to the user", *Planning Review,* May/June, pp. 16-22.

Gimson, JB. (1993), "Sun Alliance International, AD&M Systems", *IT for Marketing,* UN/COM, Brunel University 18 March, pp. 47-59.

Goodstadt, P. and Marti, R. (1990), "Quality service at NatWest Bank", *International Journal of Quality and Reliability Management,* Vol. 7 No. 4, pp. 19-28.

Hughes, T.J. (1992), "The customer database at Bristol & West Building Society", *International Journal of Bank Marketing,* Vol. 10 No. 7, pp. 77-9.

Jackson, DR. (1991), "Bingo...how Blue Cross created a breakthrough in insurance database Marketing", *Direct Marketing,* August, pp. 28-30.

Luthans, E (1991), "Improving the delivery of quality service", *Leadership and Organisation Development Journal,* Vol. 12 No. 2, pp. 3-6.

Moriarty, R.T. and Swartz, G.S. (1989), "Automation to boost sales and marketing", *Harvard Business Review,* January/February, pp. 100-8.

Morris-Lee, J. and Erickson, D. (1991), "Advertising production connects to Database Marketing", *Direct Marketing,* August, pp. 24-7.

Mueller, R.K. (1988), "The care and feeding of advisory boards", *Journal of Business Strategy,* July/August, pp. 214. (This paper is not focused on Editorial Advisory Boards but is insightful for all working with regular advisors.)

Robbins, J. (1991), "Turning the tables from service to sales", *Training and Development Journal,* April, pp. 74-6.

Roscitt, R. and Parket, LR. (1988), "Direct marketing to consumers ", *Journal of Consumer Marketing,* Vol. 5 No. 1, pp. 5-13.

Royal Mail Streamline (1993), "Rover Cars: the catalyst and conquest 1991 Direct Marketing Programmes", European Case Clearing House, Cranfield.

Schell, E.H. (1992), How to make millions with Database Marketing", *Datamation,* 1 August, pp. 77-9.

Spedding, K. (1988), "Putting the wellie in", *Direct Response,* July, pp. 56-8.

Ward, J.M. (1987), "Integrating information systems into business strategies", *Journal of Long Range Planning,* Vol. 20 No. 3, pp. 19-29.

Wills, G. Bruce, B. and Duncan, T. (1991), "Creating a marketing intelligentsia", *Marketing Intelligence & Planning,* Vol. 9 No. 4, pp. 1-20 (see Chapter 11).

Wills, G. (1992), "Enabling managerial growth and ownership succession", *Management Decision,* Vol. 30 No. 1, pp. 10-26. (Describes the previous programmes in 1991 which, among other outputs triggered the Literati club itself.)

Wills, G. and Wills, J. (1992), "Journey to marketing clubland", *Marketing Intelligence & Planning,* Vol. 10 No. 2, pp. 22-36 (see Chapter 12).

Chapter 14

Realizing the benefits of a marketing intelligentsia

All good tales have to have a beginning. This one is no exception to the rule that after a while their beginning is simplified and becomes apocryphal.

In 1987 our marketing services executives queued for a month to obtain address lists of existing subscribers, held on a product basis. They were chastised for disrupting the real business of subscription order processing. They were chastised for disrupting cash flows if they promoted late. To tell the truth it was far easier to rent a list that looked good and blast away at that.

By 1994 many of the same individuals had designed, developed and implemented a £1 million integrated customer environment (ICE for short). They were among the most "intelligent" marketing professionals in the publishing industry anywhere in the world. They have achieved this not by buying-in a re-engineered approach to their tasks but by probing and action learning their way forward. Each and every phase of development has had a precise commercial focus to provide a metric of success or failure. They would not now entertain buying-in a list until they have exhausted all their best ideas on how to develop their existing customers.

ICE embraces customer development processes, subscription services and the collections of financial services within MCB University Press. But the self-confidence it has brought to the company's marketing intelligentsia has had impacts across the whole enterprise. The acquisitions team has evolved research and evaluation criteria that build on the lessons learned by the marketing intelligentsia; the editorial quality initiative championed by Publishing Logistics is building increasingly on theories of authorship evolved by the marketing intelligentsia; and the enterprise is addressing the challenges to its future posed by electronic publishing using the prototyping approaches pioneered by the marketing intelligentsia over the past seven years.

Such proud but well-substantiated assertions have not come about by accident. They have come about because of the strategies deliberately pursued at MCB University Press.

There has never been a "Big Bang" approach. There has never been speculative investment in new technology. Rather there has been a determination always to be second wave with our technology – to be applications re-engineers, not blue skies investors. We watch, we listen to the experiences of others, we monitor changes in our marketplace and then we seek to identify how on earth such notions, such goings on, can possibly be made to fit what we currently understand we can accomplish well. Such an approach has, for example, yielded *en passant* a document delivery service at half the price and four times the speed of the British Lending Library in our areas of publishing expertise.

With Bev Bruce and Tracy Jordan,
Marketing Intelligence & Planning
Vol. 12 No. 6, 1994

The approach can clearly be seen as proactive with the technological futures so regularly unveiled and discussed, but always with the focus on how can we cross the ground commercially between where we are and where it is suggested we could be. Technological forecasters have long described it as a norm-ex approach – reconciling the normative and the extrapolative. We remain keenly aware of its potential shortcomings as pointed out by the proverbial Irishman who, when asked how to get to the normative outcome, ventured the opinion that he would not start from here, i.e. he would not concur with extrapolation. But equally we are aware of its strengths if we are to manage the risk profile of organic growth.

The purpose of this article is not to boast about the way in which we have set about creating a marketing intelligentsia and evolved its skills in marketing and promotion first to our existing customers and then to profiled prospects. It is to share what we have done, and where it has succeeded well and where it has failed. In doing so, we shall most especially dwell on two vital realization factors:

(1) the realization of what our marketing intelligentsia could offer to spur the corporate vision; and

(2) the realization of a handsome return on the financial investment we have necessarily made.

From our continuous review of the literature in this area for a decade we believe such analysis from participant observers may go some modest way to reassure the remaining doubting Thomases but a long way towards creating a much more assertive, adamant role for marketing professionals as we approach the millennium. The marketing intelligentsia, as we can now understand it, has indeed got a seminal role to play in the twenty-first century that embraces far more than the vogue accountant of the 1990s had to offer.

Defining a marketing intelligentsia

A marketing intelligentsia is a team of professionals who have taken the time and care to use database marketing technologies and the analytical procedures that have evolved with them to become highly "intelligent" about how to serve their existing customers most profitably. We have described before (Wills *et al.*, 1991) why we were confident, and how it could be achieved. Nothing in practice or the literature since has contradicted that view – indeed the literature continues to be wholly supportive (Cobb, 1991; Edelman and Silverstein, 1993; Shergill, 1993).

The opportunities for greater effectiveness and efficiency are legion. They encompass:

- segmenting markets and customers;
- focusing product development and enhancement;
- interactive researching of customers on a wide range of germane issues;
- targeting of field sales representatives;

- profiling of prospects for direct promotion;

- tailoring of offers throughout comprehended life cycles.

While it might be argued that this is no more than a logical extension of what the marketing research industry has offered since the 1930s, there is of course a vital difference – a true discontinuity. The intelligence we derive from database applications is about real customers, known individuals. It is a census not a survey. It is a pathway to an interpersonal business relationship, with an individual we will meet only if it makes commercial sense to either one or other of us – rather like a pen-friend.

Buckingham (1992) describes, for example, how Nielsen is building linkages between its legendary retail and consumer panel data and the retailers' own product and club data banks. A channel partnership has been forged with Safeway. Jansen (1992) reports how Elizabeth Arden has built a database of over 200,000 customers at retail counters taking that enterprise into the home as well as the department store. These developments complement efficient consumer response (ECR) logistics systems linking manufacturers and retailers (Wood, 1993).

In many ways the most wholesome and exciting thing about it is our new-found ability to address profitably the waste in most promotion and advertising and to avoid the rightfully criticized offence of environmental communications pollution levelled against less intelligently driven marketing activities which Kitchen (1994) intriguingly characterizes as a Leviathan, after Hobbes. It is perceivable that this can be accomplished well ahead of legislation on the rights of customers to be spared unsolicited communications in North American regulations and European Union Directives. (Cespedes and Smith, 1993; Coad et al., 1992). It is perceivable also that it can avoid further communications super-taxes as already collected on TV advertising in several countries notionally designed to increase the cost of wasteful activity and thereby accelerate the quest for greater effectiveness.

Let us conclude these definitional remarks on a marketing intelligentsia by asserting that this rather precious description is our determined attempt not to dub it simply "database marketing". Just as tandoori precisely describes a method of Indian cooking, so database marketing accurately describes the way we electronically handle our data on our customers. But database marketing is as woefully unsatisfactory a description of what we seek to achieve as tandoori is of the dining occasion! Database marketing is neither the beginning nor the end of the description of our marketing intelligentsia – it is simply the way we capture and process our data. Intelligence in: intelligence out is the challenge. Our primary concern is therefore not the latest technology, but the quality of our own creative marketing imagination and our organizational effectiveness in delivering the benefits. Haynes et al. (1992) hinted at what we imply when describing a value-added customer database that seeks to encompass customer-initiated contact. They further cited Feldman's (1988) insistence that organization culture must be aligned with and accepting of the intelligence received if realization is to occur. Fletcher *et al.* (1992) go further to emphasize that the organization structure must be aligned too, and the customer database viewed as a corporate asset. As Woods (1993) argues, this latter theme takes a great deal more than the investment in computer hardware and software on to the balance-sheet.

Small steps ensure long journeys

Bloomingdale's, one of the leading US retailers, argues for small steps as the way to make long journeys (Smith, J., 1992). Fletcher *et al.* (1992) say most experts agree. Only by taking small steps and evaluating them can the realization be accomplished by the organization and the learning internalized for future use. And so it was at MCB University Press. We had no capacity, or capability, as a small academic publisher to think big but we were determined enough not to allow either our own IT staff or outside consultants to tell us what to do. We were stubbornly of the view, and remain so to this day, that we had to learn and learn again what we could achieve and only when we knew could we make effective use of consultants or IT techies (see Rogers, 1989). We overturned substantive schemes from able consultants; we changed our senior IT manager ... we grew our own staff first as gifted amateurs, now most considerable professionals.

Our original thought processes for cross-selling in MCB University Press go back to our inception as a company in the late 1960s. We printed insert leaflets and placed them on a regular basis in "compatible" journals; and we produced monographs and conference reports in allied fields, using leaflet inserts as seemed sensible. We even bought in other publishers' remaindered stocks in our own fields of interest. Such elementary strategies worked well as was to be expected. They were simple to achieve too because we had only half-a-dozen serials. In 1994 we have 130.

Nonetheless, despite such activities taking place, since our lists were small, the major push was to rent lists and promote to the world. Today's regional profile shows half our subscribers live outside the UK and produce nearly two-thirds of our revenues – no bad outcome.

Cross-selling eventually became a focused as opposed to a casual objective under another name – the thickening-up campaign or TUC.

While TUC was able with the then database to cross-sell subscribers from one product to a similar field of interest after extensive manual deduplication, we were unable without much massive manual intervention to cross-sell on industry- or job-specific selections. Since we know both of these to be highly promising approaches, it was readily apparent that we should find ways forward as quickly as possible. It was also abundantly clear that the extant database was all too often not in possession of all the data we needed on industry or job responsibilities of our subscribers.

In 1988 a taskforce was established from among potential users, i.e. currently frustrated marketing services executives. This ethos took the responsibility away from the subscriptions-procuring and IT departments to the marketing staffs themselves, who were, let it be said, no computer experts. But they did know what they wanted; and they had the support of an outstandingly innovative board member – not, it should be added, the board level marketing experts who always remain (sincerely and rightly) sceptical about the potential cost/benefits when blue skies projects arrive.

Such cost/benefit anxieties have constantly been pleasantly unfounded, although there is always a temptation to believe their articulation helped everyone keep their feet ever more firmly on the ground.

Whysoever, what emerged from the taskforce was an exemplary and immediately successful outcome. "Operation profile" as it was known found an extremely marketing user-friendly database which took downloaded data from our subscriber database on a monthly cycle and was then well able to manipulate it as required. MCB University Press's success at this stage in creating its first marketing database is echoed in the achievements at AT&T reported by both Schell (1992) and Heller (1994), and in the UK for Allied Irish Bank (Hughes, 1992; also see Madden, 1991).

From profiles to relationships
The downloaded marketing database was, naturally, welcomed by those who had worked long and hard on the taskforce of "Operation profile". They were quick to put it into action to achieve the goals they had originally set for themselves. To help the remaining marketing services executives also to realize the benefits a *Marketing Database Manual*, with detailed campaigns, was published, with teach-ins and a formal launch occasion. Despite such efforts, the first year saw only limited use of the database. Every campaign conducted was monitored and the results showed all the anticipated benefits, but not enough was being done. The Board resolved, as a result of its awakening interest and quest for a return on investment, that for year two no prospecting campaigns would be approved until each and every marketing services executive had demonstrated all possible marketing database campaigns had been given first priority. This decision was a vital watershed in changing the attitudes of MCB University Press at large towards databased cross-selling but it very soon reached its limits. Once dutifully done, the real work of marketing services executives still remained the renting of good lists although … there were seemingly never enough good ones to be found. The Board's willingness to invest in promotional campaigns was seldom matched by sufficient action.

So, a brainstorming session was held in Spring 1991 with the ambitious title: "How can we spend another £1 million on promotion next year?" Extrapolative suggestions for finding more list brokers not just from Europe but worldwide arose. But most significantly of all for the realization of a marketing intelligentsia was the suggestion (from a non-marketer) that we should form a club among our two most significant customer groups – our librarians and our human resource professionals. The initial reasoning was to avoid "annoying" them by repetitive mailings in fields not of direct concern within their broad areas of interest. On deeper discussion it was obviously also clear that a great deal more attention and focus could justifiably be given when we did know what subscribers' true areas of interest were.

We have described in detail (Wills and Wills, 1992, 1993) how we then journeyed to marketing clubland and what we believe we learned there. Postal and telephone and field visit questionnaire approaches were differentially employed in the UK, USA, Japan and elsewhere among all existing subscribers and a greatly enhanced approach to data collection for new arrivals was adopted. Private media were gradually evolved (Cross and Smith, 1992; Minett, 1993) that further enabled us to soft-sell our products but more importantly develop our relationship with our customers as club members.

It went well beyond our journals *per se* to a wider framework of interest, concern and ultimately to loyalty wherever that might mean. We reviewed, and re-reviewed, the emerging literature – conveniently available from our own incomparable ANBAR Management Abstracting services. British Waterways (1993), Gattuso (1993), Morrall (1992), Rowney (1993), Sambrook (1993), Shani and Chalasani (1992), Sparks (1993), and Ullmann and Wonnemann (1994) all offered further insights for our thinking and action programmes from the reports of their club approaches in Swatch watches, credit cards, packaged goods, warehousing, hotels, leisure and expectant/new mothers. (The fact that few of these articles attempt anything more than to tell their story in no way detracts from their professional and comparative academic worth. They supplement and extend the extensive literature already reported in Wills and Wills (1992, 1993).)

The establishment of two clubs at the outset was perceived at the time as somewhat ambitious. As it transpired, it was extremely helpful. They represented not only different professional groupings but of course quite different customers. Librarians were predominantly acting as brokers for others across their institutions – faculty members, students or managers. Human resource professionals were predominantly purchasing on their own direct behalf. The first club had the stability of the library as an institution behind its members – and a continuity of subscribing behaviour beyond the needs and wants of a single individual. The human resource professional was highly mobile as the fortunes of his enterprise, his career and his profession ebbed and flowed.

Industrial relations or training as key issues have, for example, been displaced in a decade by learning organizations and quality assurance. Organizational restructuring in the recent world recession has seen a major drift from full-time human resource corporate staffing to bought-in consultancies – the latter being quite different purchasers of academic and professional literature from their corporate predecessors of the 1970s and 1980s.

Thinking laterally in the enterprise
The potential benefits of club thinking, the creation and maintenance of relationships with key individuals outside the enterprise itself, had a

fascinating spin-off. A range of action learning management development programmes produced *inter alia* a proposal that the authors who wrote for MCB University Press should be constituted as a club too. The senior manager responsible for the Librarians' Club – known as Library Link – took charge of the authors' club, known since as Literati Club. As will be recounted later, this development was to be a major fillip to a widespread Quality Initiative throughout the area of editorial procurement and publishing. At this juncture, its most significant difference from the two marketing clubs was that the Literati Club database was prototyped completely outside the head office, using dedicated, customized software systems. While this again posed familiar problems for adoption among those who did not own the system, it had the important impact of creating dissatisfaction with the recently arrived marketing database. Literati Club was a more flexible and readily accessible system and had been developed with a wider range of applications in mind for the author and editor relationship.

Yet at this moment, the Board resolved not to proceed to either a more sophisticated system or indeed to an integrated customer system. It shifted the enterprise's priorities away from enhancing customer focus to a major strategic challenge on the manufacturing side of the business – to the ever-looming challenge posed by electronic publishing. While it was in no way possible to perceive what the future would hold in store for a publisher such as MCB University Press it seemed clear that the manner in which the knowledge/articles were held should itself become a database, accessible to customers in whichever way was most convenient – perhaps by keywords or abstracts rather than as full-text journals on a subscription basis. As such the Board took the decision to re-equip the Production Department of the enterprise so that such electronic technologies would be in place by late 1993.

It takes only a moment's reflection to see that the re-engineering of the way the enterprise viewed the production process was wholly compatible with the way the marketing intelligentsia viewed its customers – and, through the intervention of the marketing intelligentsia, the way we also viewed our authors in the Literati Club.

This move into MK 3 electronic production systems was the outcome of the same action learning management development programmes as had triggered the Literati Club. The Board member who had first championed Operation Profile and the marketing database had widened her concern with information management across the whole enterprise. With an IT taskforce she had developed a schema of the information systems of the total enterprise and given the component elements to a series of senior and middle management teams to explore. Their action learning reports (Wills, 1992) became, as they were implemented, the platform for extensive empowerment of the rising generation of senior and middle managers. The Board crafted systems that let the managers lead (Stayer, 1990) but went further by inviting the managers themselves to join the specifying taskforces.

Two other information management projects were activated – both to improve the logistical flows of the enterprise. The first, where the ownership was within the taskforce which created a promotion information system (PIS), has worked well. The second, a production-tracking system (PTS), took nearly two years to come online. Despite a perceived need from outside production that it was appropriate, it failed the test by having been conceived outwith the department – a traditional enough lesson.

Determined to compose not simply to play
During this lateral thinking phase, marketing services executives and the minders of the marketing database known as the marketing development unit (MDU) were not idle. They were getting used to the idea of using the database in a host of ways, and the clubs were growing in self-confidence as their profitability was established beyond reasonable doubt.

Once the commitment to production, and to PIS and PTS, had been made, it was clearly time to advance the marketing intelligentsia's contribution. At this moment we got it wrong. A cumbersome customerization taskforce sat for months on end made up of representatives of all the implicated parties – rather than committed intrapreneurs. It concluded laboriously as we should have predicted (Wills, 1991) that customerization was a good idea and left it at that. The hands-on intelligentsia were swiftly recalled and proceeded immediately to get all their colleagues to compile their "wish list" for the next generation and proceed at the express request of the Board to prepare their own specification for what they wanted. Armed with that, external consultants were invited to offer their proposals to help us move forward.

Incredibly as it may now seem, they urged us to re-engineer our enterprise to suit their view of how we should operate rather than seeking to build and professionalize what we had already envisaged. What we actually believed we required was guidance on how technology application could exemplify our marketing and information orientations (OASIS, 1991). We resolved to do it ourselves, with a strong new appointee as IT manager bringing implementation expertise to MCB University Press from outside to assist us.

Yet we did re-engineer our enterprise – in December 1992 – in order to create a vision for 1994 that would consolidate the learning we had achieved as our information management strategy had unfolded (Cooper and Stephan, 1994). The responsibilities for comprehending the flow of authors' contributions from their minds and PCs to eventual retrieval and readership across the world was integrated under the banner of publishing logistics. The business of developing our profitable relationships with the customers we had was segmented out from the process of prospecting. The Board's dictum that no promotional expenditure should go on prospecting until every possible expense had been invested in the customers we already had was now enshrined in the organizational structure.

The structure for the future exploitation of ICE, arriving August 1994, was deliberately put in place in November 1993, still making the best use it was able of the downloaded marketing database – but sharing in the specification, as it went along, of the ICE systems outcomes (Figure 1). Furthermore, so intent was the Board to ensure ICE really was what was needed, the period between the announced reorganization in December 1992 and its implementation in November 1993 was devoted to the staff involved deciding how they wanted to organize themselves within the new structure. OASIS (1991) describe this as having reached the Composer phase in the evolutionary progression of sophistication in managing marketing information and we must modestly agree. This would seem to position us as champions of end-user computing. Certainly we must agree with Runtagh (1992) that information and technology are powerful cultural change agents if they are well managed.

Yet we still firmly believe the arrival of ICE as an integrated system will not for one moment limit the continuation of prototyping outside the main integrated system. Nor do we believe, as OASIS suggests, that this puts us at a competitive disadvantage. The dynamics of a living and evolving marketing intelligentsia demand that the composers have licence to go away on their own, with their own unproven priorities, to evolve and develop ideas – perhaps for the next integrated system or perhaps to wither quietly away. As such PIS will be almost wholly integrated within ICE three years after its prototyping. So too will the Literati Club. Yet already on the outer edges of our creative marketing thinking new prototype systems are being dreamt of and authorized.

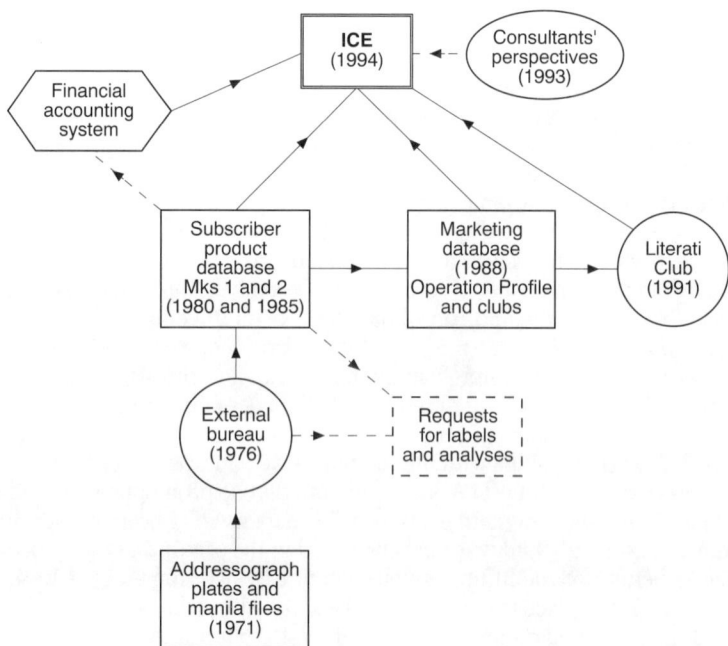

Figure 1.
The emergence of ICE –
our integrated customer
environment

Intelligently waiting for ICE

Waiting for ICE has not been as depressing as it might sound. It will be recalled it was heralded at the end of 1992 and was only six months late arriving – in August rather than March 1994. The fact that we all knew it was coming, and the wish-listing that preceded it, made us all think as hard as we could about the intelligent ways of behaving and relating which we wanted to capture easily with the new software. The fact it was grindingly tedious to stroke and flex the data in the extant marketing database was not seen as a drudge – it was the way to know what we wanted. As such the highest priority was given to modelling our customer relationships and to capturing data with integrity to allow it to happen conveniently. To this end the concept of the subscriptions data capture phase had to be enriched most considerably.

This was aided by the appointment to head that department of the former Human Resource Professionals Network Club Convener. The linkages with Financial Services also had to be modelled and built – and again the head of that department was a former head of both subscriptions and marketing services.

Within Customer Development, now managed as two major profit centres, the early models of the two Clubs were dramatically extended (Figure 2).

The largest number of MCB University Press customers are individuals, normally professional managers, buying with their company budget over which they have discretionary control. Their purchases are typically driven by a desire to extend their current level of professional expertise and, as their careers develop, those desires change and evolve. This motivation to subscribe to a particular journal at any given moment in time inevitably contains the seeds of its own cancellation in due course. Indeed, the cancellation rates for such customers are 50 per cent higher than for librarians. This does not surprise nor alarm us but it puts an especial premium on building a relationship with such customers that enables us to move with their evolving career needs – and/or with their job changes and the names and interests of their successors.

As such it will be perceived that what is needed is not only the life cycle of a given professional subscriber in-post but also the longitudinal career cycle of that subscriber too.

Hansotia (1993) shows how, by forecasting when your customers will leave, you can form strategies in advance to evolve the relationship more effectively.

We also need to comprehend not only such individuals' own interests but also their spheres of influence. A logistics or finance manager has less likelihood of influencing the knowledge-gathering or journal-reading habits of colleagues than a human resource manager engaged either in training or staff appraisal/career guidance work. Consultants are far more likely to need access to the literature across a wide range of knowledge as they go from project to project, or industry to industry, than an individual corporate manager in a specific role. For this latter comparison, speed of access may well be vital to a consultant whereas a corporate manager may be more appropriately characterized as browsing/staying up to date. At different ages, and at different phases in any given roles, our professional

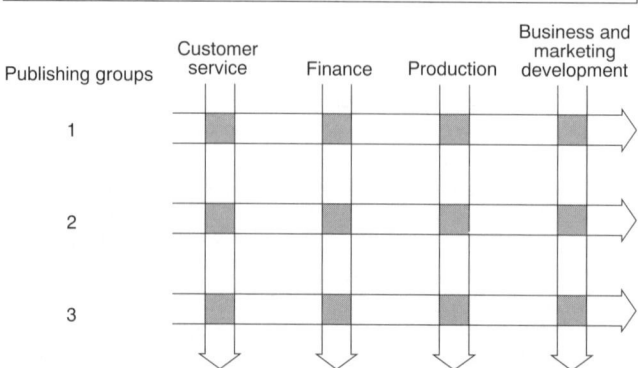

Previous organization — **Product-oriented**

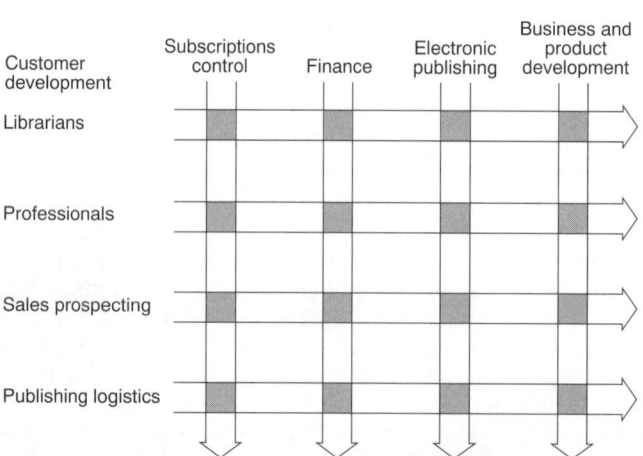

Revised organization — **Customer-focused**

Figure 2.
How the development of
a marketing
intelligentsia changed
the organization
structure

customers will have totally different requirements of their journals and information services. They will have different lifestyles too (Dwek, 1991).

Our challenge is how to collect this information, or identify surrogates for it, and how then to satisfy such diverse segments cost-effectively across some 130 journals in some 15 product groups with some 40,000 customers. This latter point, despite a relatively high average sales value and repeat purchase probability, is vital when considering the cost effectiveness available. Some discerned segments may contain no more than a dozen customers at a time for a focused relationship. Can it be worthwhile commercially?

Such philosophizing has led, naturally, to a series of specific, frequently automated campaigns, which are constantly evaluated and replaced/enhanced.

These are illustrated in Figures 3 and 4 for first-year subscribers (who have the greatest propensity to cancel) and for the global renewals and associated campaigns throughout the customer life cycle. As many as 50 per cent of first-year professional subscribers will not renew unless listening and nurturing campaigns are conducted (see Williams (1990) for discussion of second purchase levels among mail order customers at large and Gray (1993) on the use of billing opportunities with credit card to sustain and enhance the sales relationship).

The greatest continuing challenge, once the first renewal has been achieved, is to sustain the relationship notwithstanding career progression and job changes. The good news is that as each professional moves on there is a high probability another takes up the role – but of course frequently with different characteristics to be discovered and catered for. This requirement to gather continuous intelligence throughout the life cycle of a customer, rather than simply at the outset and at cancellation time when (heaven forbid) it is almost always too late, requires determined effort; and considerable investment. The development of technology and in particular of ICE can now manage such campaigns (see Cobb, 1993).

Stage 1	Prospecting and Help Desk

Ensure that promotional mailings accurately reflect journal content *and* present in most persuasive context for the targeted prospect, e.g. region, industry, psychographics. Help Desk available and "answers" captured in ICE

Stage 2	Order acknowledgement

Ensure customer order and payment received and when first issue will arrive

Stage 3	Welcome wagon

Welcome as a new reader, ensure profile data are complete and alert to other matching MCB products

Stage 4	Nurture first-year subscribers

Sustain contact throughout the first year seeking comments on satisfaction/dissatisfaction with the journal purchased and any profile changes

Stage 5	First renewal

Offer focused consideration at renewal time that affirms interest and alerts reader to following year's service

Note:

This nurture process is appropriately conducted with all subscribers after their first year on an automated basis

Figure 3.
First-year prospecting
campaigns

End of volume

Renewal campaign
(eight-stage chaser)

Acknowledgement that their subscription is about to expire and affording the opportunity to prevent an interrupted supply. Informing the customer of the added values and enhancements to the new volumes to ensure their continued custom

Continuous campaigning

X-selling campaign

Recognition that MCB products are compatible with those they currently take

Industry campaign

Recognition of the industry in which they work

Job function campaign

Recognition of the job function they have

Region campaign

Appreciation of the region of the world in which they live and its language

New appointments campaign

Recognition by MCB that they have been promoted within their organization to a new role and that they may well have a successor

Mobility campaign

Recognition that MCB does not want to lose their custom and that they may still need MCB products in their new role and that they may well have a successor

Cancellation campaign

Assurance that MCB does want constructive feedback to regulate products and services if necessary, that it does not wish to lose their custom and that it is seeking to find an alternative product to fit their needs.
10 Point Action Plan — what can MCB do that will fulfil their requirements or ensure that it has taken on board their feedback, e.g. where they do not have enough time to read journals offer retrieval/abstracting alternative

Capture customer feedback

Lapsed campaign

Recognition that MCB may have new launches or acquisitions that may interest them and resurrect their interest in its product portfolio

Stage 1
First forward chase
prior to expiry

Stage 2
Second forward chase
prior to expiry
Ensure no forward chasing of new launches
Recognition that they have yet to receive the complete volume before they can make the decision to renew for a second time

Stage 3
Renewal notice despatched
with last issue of journal

Stage 4
First post chase after expiry
Reiteration of the journal's attributes and reminder to the customer of the necessity to continue subscribing, together with highlights of what is to come

Stage 5
Second post chase
after expiry
Reiteration of benefits and incentivization

Stage 6
Telephone renewals

A personal and regional voice from MCB reminding them of the opportunity to renew and updating their profile while giving us views on the product and the service

Capture customer feedback

Stage 7
Retention campaign

Recognition that MCB does not want to lose their custom and providing them with an opportunity to feed back on product and service quality. Ensuring that any change in interest areas or personal data is updated and trying to find alternative products to ensure an extended lifecycle

Capture customer feedback

Stage 8
Replacement/Promotion campaign

Recognition from MCB that they may have moved on in their role, been promoted and/or replaced another in their organization

Capture customer feedback

Figure 4.
"Moments of Truth" during a professional customer's life cycle

By definition journal publishing is done at a distance from the customer and reader. Questionnaires can be, and under the club format, are despatched but as would be expected only modest levels of response (25 to 40 per cent) are achieved. The imminent arrival of the opportunity of interactive computing relationships, e.g. computer conferencing and Internet, also opens avenues that have thus far been only poorly accessible.

This is one more illustration of how our marketing intelligentsia shares in and at times leads the development of other areas of the enterprise. Such conferencing clearly has implications also for article procurement and referencing processes as well as for publication itself.

Relationships with librarians
The strategic thrust described this far to achieve more intelligent relationships with professional managers is mirrored in MCB University Press's relationships with librarians. But there are two vitally important channel distinctions to note from the outset. The great majority of librarians order their journals via agents, of which the world is dominated by some six or seven. Second, as already noted, librarians purchase only a small percentage of journals for their own professional purposes. The great majority are for the library's users needs which are determined in decision-making units (DMUs) that frequently are a long way away from the librarians' offices. So the challenge to build customer relationships with librarians has to be qualified at once by observing that success will be achieved only if the whole channel comes together effectively.

Against such an analysis it will not be surprising to discover that the management of librarians, their agents and their wider DMU in institutions was placed under a single senior manager. Nor will it be surprising that field sales visits are made on a regular basis around the world to agents. Nor that the agents provide an extremely important source of intelligence on development among librarians. What was to us surprising was the realization that we could work with those agents (not ours, remember, but the librarian's) to undertake what for us were prospecting campaigns – but to the agents were customer development. By matching our titles against compatible titles not published by ourselves but taken by a library, we could at a basic level show agents how to segment their promotional campaigns to sell our journals. We did, of course, incentivize them with excellent first-year commissions, and a better than normal commission in year 2.

Their performance on our behalf has been spectacular as shown in Table I. Not surprisingly we are greatly increasing our efforts in this direction all the while.

Interestingly, however, as we probe more deeply into electronic publishing, the agents find themselves more likely to become our partial competitors in the marketing channel. They are keenly interested in not being excluded in the future of electronic publishing and their privileged position of contact with so many of the world's libraries has afforded them the opportunity to establish, for example, document identification and/or delivery services, e.g. Blackwell's

UNCOVER. The case for further distribution channel alliances seems strong, and demands creative marketing initiatives.

| Year first sold | | | Percentage | | |
Renewals year on year	1990	1991	1992	1993	1994
1990	100	–	–	–	–
1991	72	100	–	–	–
1992	64	76	100	–	–
1993	75	82	100	100	–
1994	67	72	88	71	100
Gross income thus far as percentage of first year income	460	409	356	189	100

Table I.
Index of librarians'
agents sales
achievements for
MCB University Press

A second field for development with librarians speedily arose when we began to look intelligently at our own key customers (KCs). It emerged that less than 1 per cent of customers – all of them librarians not professional managers – accounted for 15 per cent of gross sales revenues. A senior manager, previously head of customer services, was made responsible for all aspects of their relationship with us. She gathered intelligence on their renewal patterns just in time to avoid us running a pointless campaign with meaningless incentives (Woolf, 1993)! Their renewal levels stand well above 80 per cent, and librarians' cancellation decisions will not be avoided by a direct appeal to them, since they are almost invariably taking their cue from their DMU among faculty members. To develop KCs accordingly calls for a strategy towards library users at large – and to do this they must be more broadly identified.

Library Link's club questionnaire has, for three years now, sought out this information, always asking for details of any departmental library decision groups. Once again, only limited success was achieved by this approach – or could be expected in the long run either. Supplementation by telephone interviews was more effective in North America but field visits were unavoidable for example in Japan. To achieve a more sustained breakthrough in this area and for librarians in general who were not yet KCs, we concluded that further development of our promotional mix was required. A field salesforce is being created on a joint venture basis between Literati Club, Library Link and Sales Prospecting – of which more later.

Over 90 per cent of librarian customers are not currently managed as KCs. The Library Link team uses the intelligence captured to provide sample copies of our journals in the obviously appropriate fields. The average response rate for unsolicited journal samples despatched for focused reasons ranges between 3 per cent and 8 per cent. The further intelligence we accordingly seek to gather is why the take-up is so low – albeit that it is profitable at such apparently low levels.

In addition to the obvious response that the journal itself was not what might be anticipated/value for money, we have addressed the query: are we giving our journal its best possible representation by a single issue as a sample? In the same way as a prospecting leaflet must tell the total story of the journal's benefits, so too should what we despatch by way of a sample. No one particular issue can do justice to the coverage of a full volume, and even the inclusion of an index of the previous year, or choosing the "best" of the previous volume can do the task thoroughly. Accordingly, we are now using what are termed Quality Sample Campaigns.

For our major titles only thus far we have assembled a selection of articles including recent Article of the Year winners. The Editor has prepared an in-depth explanation of what the objectives are, the Advisory Board's membership is clearly demonstrated, and abstracts and keywords for the past three years' volumes of all articles published are given. It is of course a special publication kept in print on a rolling basis, but using standing copy.

Library Link does more, however, than mail out sample copies in areas of perceived relevance. In 1991 it began to make available between four and eight weeks prior to publication the contents pages, keywords and abstracts of all forthcoming MCB University Press journals. These were mailed monthly in hard copy format covering all the areas of the librarians' interest. It was an extremely expensive campaign but greatly appreciated by librarians. It has now gone interactive electronically via Internet, offering even faster revelation of advance information, plus publication details as well – scheduled dates for despatches, frequency, etc. This development service from Library Link constitutes MCB University Press's first wholly electronic publication for its customers. It can, of course, be made available to a much wider range of prospective customers worldwide with opportunities for hard copy or electronic document delivery of any MCB University Press article – without need to subscribe to a full volume of any journal, just to the electronic service itself. By listening to the voice of our customers in our segmented marketplaces, a wave of new product development is moving in to supplement the conventional new journal launch driven models (Cooper, 1994; Johne, 1994).

Intelligent prospecting

The overwhelming proportion of MCB University Press's gross income each year arises from existing customers. First they renew subscriptions; second, they purchase additional subscriptions. As such, the formal restructuring of the enterprise to manage relationships with the customers we have is timely, even overdue. Nonetheless, much future growth and much developing profitability will arise from those who do not already have any relationship with us. The third major customer management area in MCB University Press is, accordingly, Sales Prospecting – the search for as many new customers as can be profitably found across the world. This search is obviously made more intelligent as we learn more about the profiles of the customers we already have.

Librarians, our analysis shows, are the most likely to renew, so the highest prospecting priority should be given to them. Yet they are the customers most likely to convert to electronic search and retrieval services to serve their users in the years ahead – albeit that their DMU faculty members seem still to want their own authored articles printed in hard copy journals.

So, while directing as much effort as possible to the finite number of prospect librarians worldwide, activity must also be targeted to the professional managers *per se* both within and without librarians' DMUs. Yet simply anticipating that the creation of a prospect, a first-time purchase transaction, is sufficient misses the point about building relationships for future profitable development (Axson, 1992). A promotional investment sequence in terms of annual budgeting accordingly emerged as follows, which although putting prospecting third and fourth leaves ample resources available for them:

- 1st priority – customer development with librarians and their agents;
- 2nd priority – customer development with professionals;
- 3rd priority – prospecting for librarians;
- 4th priority – prospecting for professionals.

There are two reasons why ample funds are always available – first, the sheer limitation of activities under priorities 1 and 2 brought about by their relatively limited numbers. We have no more than 8,000 librarians and 40,000 professionals as current customers. New launches and acquisitions, if they are synergetically made, do offer fresh opportunities each year, and promotional impacts are carefully monitored. Except by the development of tie-ins, however, there is a theoretical limit on what can be offered under priority areas 1 and 2 from our own product range – and we have a reluctance strategically to stray too far from the knitting in terms of current product groups except by acquisition of a suite of titles.

Tie-ins undoubtedly hold good future potential (Cross and Smith, 1992) well beyond our highly successful database marketing agreements with librarians' agents. In several product groups we are working with conference organizers, book and report publishers. We can also ourselves undertake the preparation and commissioning of such products.

The major prospecting thrust thus far, excluding the imminent move into field sales visits, has been the search for yet more effective global promotional pieces and a determined quest for regional exploitation of prospect mailing lists. As was noted earlier, over half of all sales are made outside the UK into mainland Europe, North America, Japan, Australasia, the Far East, South Africa and South America. In each of these geographical regions on the ground, offices have been established either directly or in partnerships with agents to exploit local profiled lists with local return addresses, local currency pricing and local customer service points.

Australasia was built to become our third largest market after the UK and North America, the Far East has one of the best levels of promo dollar response worldwide and Japan is responding excellently to our local office staffs.

Sales prospecting is now in a position to focus its campaigns not solely on journal titles but also towards specific, customer targets with the most profitable life cycle expectancy and/or those regions of the world where the best response levels are achievable.

It is also setting its sights on "neighbourhood selling" within the existing enterprises of the customers we have. Just as librarians' DMUs are traced and targeted, so too can the individual professional manager's colleagues. It is conceivable that corporate volume terms can be offered not just for multiple subscriptions to the same title, but also for all products going to the single enterprise address.

Such activities obviously touch on/overlap with the current role of customer development with professional managers. They call for good relationships at the senior levels for all three sales departments and a performance assessment system that is non-divisive. They share greatly in staff development initiatives and cross-company taskforces as a matter of policy. They also convene their own Global Sales Conference annually to share and compare. Additionally they are continually developing a portfolio of evaluated, successful campaigns across all three fields. Nowhere is this co-operation better illustrated than in the emerging launch of MCB University Press field sales representation campaigns – introducing a fifth channel after direct mail, agents, telephone and computer access.

Field sales representation emerges
The origins of field visits to customers and prospects lies in Literati Club, the authors' club established as a lateral thought after Library Link and the Human Resources Network had flourished. Literati Club pioneered the collection of intelligence from authors and editors on a wide range of matters – not least their attitudes towards electronic data flows for their articles. By 1993, however, it found a major part to play in support of the company-wide Quality Initiative. Authors and editors were invited to benchmark our major titles against our competitors for articles they wrote. Most significantly, of course, we did not provide the judgemental criteria. We elicited these from the authors and editors first in a preliminary study.

The conclusions are not so much confidential as of little generalizable use. Depending on the particular niche market surveyed the editor was of either extreme or little importance. Prestige, content and presentation emerged, however, as the three major fields where measurement was made. The overall conclusion was, as we discerned the obvious with great clarity, that we needed at all times to accelerate the flow of top quality articles into the editorial and refereeing system as a necessary if not sufficient condition for success.

To achieve such a high quality flow required supply side contact with our authors and editors around the world. The quality of presentation of ideas in writing was frequently the reason for extensive refereeing and sometimes indeed lost publishing opportunities. We resolved that it was well worthwhile to visit the major centres of excellence in the product groups we covered to meet all potential and actual authors to encourage, even help to show them how to submit and be successful. All this of course had to be achieved without undermining the discretion of the individuals in-post as editors. Simultaneously, editors saw some benefit in development of themselves in their roles. To this latter end Editors' Development Workshops are now regularly held for novices and the more experienced, and applications software to guide and organize editors' administrative systems and article searches has been custombuilt. Its design evolved from an audit of existing best practices developed by editors around the world. It is known as Better-Ed.

When an unbridgeable editorial development gap emerges between the team in-post and our publishing needs, a tasked team is now regularly invited to intervene to correct affairs – it is known as a "Harvey Project" after the pioneering venture of its type in the company.

As soon as the idea for such field sales visits was floated, it was immediately apparent that most of the centres of excellence to be visited were already KCs or in the Library Link Club but of lesser size. Accordingly the field visit intelligence system became an amalgam of the Literati Club, KCs and Library Link Club members – co-ordinated across the world *ab initio* by the regional development managers predominantly concerned with assisting sales prospecting to exploit regional opportunities to a maximum.

While such an alliance of interests heralds a nightmare of co-ordination of conflicting or competing objectives on any given field visit (Close, 1992), it has received a major boost from the arrival of high value added electronic CD products. And the prospects in the coming 18/24 months are for even more significant high value added services as well. Furthermore, Literati Club has resolved to develop an associate membership division worldwide, for individuals who anticipate writing for our journals but have not yet done so, and whom we would wish to do everything possible to assist. We shall offer them guidance and assistance right through the refereeing process and have commissioned our own guide for them entitled "Write Right First Time".

Table II shows the model of relationships to be addressed by the field visit programme over the next three years. Every one of the significant centres of excellence for our product groups is now allocated to one or other of the relationships set down; and for each relationship field visit resourcing is being prepared (Koepcke, 1994). With some 2,500 centres to visit over the coming three years at some £500 per call, it is clear that well over £1 million will be invested. That investment must generate a goodly return in the form of a combination of publishable high quality articles and long-term sales revenues. We are all well familiar with the cost/benefits of a new subscription but the cost/benefit of new authors writing for MCB University Press we shall have to learn to understand. We know the current average cost of capturing a good quality article but we are not too certain how to measure its benefits to us as a publishing house.

	Existing Literati authors	Potential authors for MCB-UP
Centre for Excellence who are KCs	Matrimony Strategy 1	Engagement Strategy 2
Prospective KC[a]	Courtship Strategy 3	Dating Strategy 4

Note: [a]We define prospective KCs differently in this model to include all significant centres of excellence where we currently have no sales or where the sales are so modest as to constitute very considerable potential

Table II.
Field visit relationship strategies

There are, of course, also many centres of excellence and libraries outside the 2,500 we are scheduling to visit over the next three years. They will not be ignored; much more substantial contact than hitherto will be made with them, and also prior to our field visits to those as well. Contacts will be planned at conferences where they can be expected to attend, and at invited workshops to which they will make their way at their own expense. And we shall of course sustain our direct mail activities and our use of interactive computer conferencing and networking – this latter in further development of our advanced contents pages, keywords and abstracts available on Internet (Owen, 1991; Smith, 1992).

The strategies towards the centres of excellence can be characterized as follows:

- Strategy 1: *Matrimony* where a centre is both a key customer and has Literati members. We have to work hard at this relationship to keep it alive; some people take it for granted and then wonder why the partner drifts away… research shows that most people leave a database not because they were pushed or swept away, but because they were neglected.

- Strategy 2: *Engagement* where we have a key customer but no Literati Club author yet. The librarian has a relationship with us but not the faculty. We have to influence the right people to write for us and in doing so improve our quality and keep the interest going towards sales.

- Strategy 3: *Courtship* where we have a Literati Club member but the centre as yet is not a key customer. This is a fine relationship with the Literati member of the DMU able to influence the librarian to purchase and colleagues to write.

- Strategy 4: *Dating* where we have nowhere obvious to start and will need to introduce Literati events and sales activities of a highly focused nature. This is likely to be the largest target prospect group of all, with the lowest response rates, but its delineation is half the journey to achieve our goals.

Intelligently approaching electronic publishing

It seems at first glance presumptuous to suggest that one of the major benefits yet to be realized from our marketing intelligentsia is a coherent way forward in the face of all the current uncertainties for electronic publishing. Yet on reflection it is a valid suggestion. Several not insignificant elements have already been noted. It was in the field of librarian customer development that the first wholly electronic serial has been produced. And it is here that electronic conferencing has been pioneered within our educational services, known as the Atherton Intelligence System (or AIS), which offers open-ended access to full text document delivery arising from Bulletin Board Systems. Our Docutech Systems are already on the threshold of integration with fax machines for automated image transmissions (Sanderson, 1992).

The advent of the concept of a field visits campaign over the next three years has focused on the opportunities for higher value services to be offered, that can include the cost of their installation and help lines. Linkages with agents worldwide have given access to and understanding of knowledge systems which are far more comprehensive than those at MCB University Press (Reid, 1992).

From all this comes the realization that the future success of MCB University Press depends on bringing emerging technologies for knowledge interchange, capture, transformation, transmission, search and retrieval to our existing and prospective customers profitably. While the highest possible competence in the technologies will be required, we have forsworn achieving this by technological first-wave activities because of the high level risks involved for an enterprise our size. Accordingly, the only way to our future must be through an understanding of our customers' needs and wants. A marketing intelligentsia is the best way we know how to achieve that.

Over the next five years this will require a range of probing offers for our customers using electronic technologies to the full, either alone or in tandem with hard copy publishing. They must, however, be designed from the customer backwards into our enterprise and not the other way around. Our intelligence via Literati Club on the supply side of our market and via existing customers among librarians and professional managers is vital.

The realizations that we have

The development of our marketing intelligentsia has been made possible by the advent of database technologies. It has been a profitable, managed risk development because we have moved forward on the basis of specific focused projects, permitting prototyping of diverse applications to gain the benefits of intrapreneurship.

The core systems applications have accordingly developed slowly and, in computing technology, are unspectacular. However, they have afforded the most widespread adoption and ownership and, as such, the appetite for continuous further development.

Most significantly of all, our marketing intelligentsia has probed forward into the area of greatest marketing uncertainty for the enterprise – electronic publishing over the next decade – and the multiple channels that are emerging for collation, transmission and retrieval. As such some realistic, potentially profitable futures can now be dimly discerned for the transition of our enterprise from one fundamental technological capability (print) to another (electronic interchange). We most surely have not seen the end of print in our lifetime but its place in the knowledge markets of the world will be radically different in the twenty-first century. The quest is to conceive and market profitable products and services using either or both technological capabilities to the full. We believe that having a listening, interactive relationship with our customers and suppliers is the only way to know what we should seek to achieve. And it is the only way because at present neither we nor our suppliers or customers know what we all truly want. We are actually learning together (Easton and Araujo, 1994). To stay profitable at MCB University Press we must be ready and flexibly able to respond as the future comes into sharper focus.

References

Axson, D.A.J. (1992), "A return to managing customer relationships", *International Journal of Bank Marketing,* Vol. 10 No. 1, pp. 30-35.

British Waterways (1993), "Messing about on the river", *Direct Response* (UK), April, pp. 22-4.

Buckingham, C. (1992), "New capabilities: the revolution in continuous marketing research", *Admap,* October, pp. 32-5.

Cespedes, F.V. and Smith, H.J. (1993), "Database marketing: new rules for policy and practice", *Sloan Management Review,* Vol. 34 No. 4, pp. 7-21.

Close, W.S. (1992), "A software diet to fatten sales", *Sales and Marketing Management* (USA), October, pp. 76-80.

Coad, T. *et al.* (1992), "How well are we fighting the EC Directive?", *Direct Response* (UK), March, pp. 12-13.

Cobb, R. (1991), "Preparing for the Halcyon days", *Marketing* (UK), 7 November, pp. 27-8.

Cobb, R. (1993), "Technology takes the strain", *Marketing* (UK), 21 October, pp. 33-6, which examines the use of database approaches to address the challenges of small customers in the grocery, off-licence, tobacconist and newsagent sectors.

Cooper, J. and Stephan, R. (1994), "Reinventing logistics: is business process re-engineering the answer?", *Logistics Information Management,* Vol. 7 No. 2, pp. 39-41.

Cooper, R. (1994), "New products: the factors that drive success", *International Marketing Review,* Vol. 11 No. 1, pp. 60-76.

Cross, R. and Smith, J. (1992), "Staying in touch with private media", *Direct Marketing* (USA), June, pp. 28-32.

Dwek, R. (1991), "Matters of life and data", *Marketing* (UK), 7 November, pp. 31-2.

Easton, G. and Araujo, L. (1994), "Market exchange, social structures and time", *European Journal of Marketing,* Vol. 28 No. 3, pp. 72-84.

Edelman, D.C. and Silverstein, M. (1993), "Up close and personal", *Journal of Business Strategy,* July/August, Vol. 14 No. 4, pp. 23-31.

Feldman, S.P. (1988), "How organisation culture can affect innovation", *Organizational Dynamics,* Summer, pp. 57-68.

Fletcher, K., Wheeler, C. and Wright, J. (1992), "Success in database marketing: some crucial factors", *Marketing Intelligence & Planning,* Vol. 10 No. 6, pp. 18-23.

Gattuso, G. (1993), "Taking care of business", *Direct Marketing* (USA), February, pp. 20-2.

Gray, D. (1993), "Selling while billing", *Credit Card Management* (USA), March, pp. 67-9.

Hansotia, B.J. (1993), "Can you predict when your customers will leave?", *Direct Marketing* (USA), March, pp. 24-6.

Haynes, P.J., Helms, M.M. and Casavant, A.R. (1992), "Creating a value-added database", *Marketing Intelligence & Planning,* Vol. 10 No. 1, pp. 16-20.

Heller, J. (1994), "Courting customers and employees", *Total Quality Review,* January/February, pp. 15-18.

Hughes, H. (1992), "Allied Irish Bank – stealing a march on its competitors", *Direct Response* (UK), October, pp. 42-3.

Jansen, M. (1992), "Elizabeth Arden: the sweet smell of success", *Direct Response* (UK), April, pp. 39-40.

Johne, A. (1994), "Listening to the voice of the market", *International Marketing Review,* Vol. 11 No. 1, pp. 47-59.

Kitchen, P.J. (1994), "The marketing communications revolution: a Leviathan unveiled?", *Marketing Intelligence & Planning,* Vol. 12 No. 2, pp. 19-25.

Koepcke, K.B. (1994), "Automating your salesforce", *Small Business Reports,* March, pp. 14-17.

Madden, D. (1991), "Banking on a database", *Financial Times,* 17 December, p. 30, which describes the progress NatWest was making towards a new customer relationship system.

Minett, M. (1993), "Company newsletters as a sales lead generator", *Direct Response* (UK), April, pp. 30-1.

Morrall, K. (1992), "Managing the deluge of data", *Credit Card Management* (USA), March, pp. 14-17.

OASIS in association with the Chartered Institute of Marketing (1991), "The management of marketing information", from OASIS, Tectonic Place, Holyport Road, Maidenhead, Berkshire SL6 2ET.

Owen, D. (1991), "Microsegmentation's role in maximising representatives' performance", *Business Marketing Digest,* Vol. 16 No. 4, pp. 65-70.

Reid, C.D. (1992), "Doing business in Europe", *Marketing Intelligence & Planning,* Vol. 10 No. 6, pp. 31-6.

Rogers, M. (1989), "Creating the marketing receptive environment: overcoming the two-year hatchet limit for a firm's first marketing director", *Journal of Business and Industrial Marketing,* Vol. 4 No. 2, pp. 17-25.

Rowney, P. (1993), "Buy, buy, baby", *Direct Response* (UK), October, pp. 52-4.

Runtagh, H. (1992), "Information and technology: cultural change agent", *Journal of End Use Computing,* Vol. 4 No. 4, pp. 37-41.

Sambrook, C. (1993), "The world's biggest hotel database", *Marketing* (UK), 4 March, p. 19.

Sanderson, D. (1992), "Telephone technology that could change direct marketing", *Direct Response* (UK), February, pp. 30-1.

Schell, E.H. (1992), "How to make millions with database marketing", *Datamation,* August, pp. 77-80.

Shani, D. and Chalasani, S. (1992), "Exploiting niches using relationship marketing", *Journal of Services Marketing,* Vol. 6 No. 4, pp. 43-52.

Shergill, S. (1993), "The changing US media and marketing environment", *International Journal of Advertising,* Vol. 12, pp. 95-115.

Smith, F. (1992), "Long journeys start with small steps", *Direct Marketing* (USA), September, pp. 26-31.

Smith, J. (1992), "Channel wars", *Direct Marketing* (USA), April, pp. 33-7.

Sparks, B. (1993), "Guest history: is it being utilised?", *International Journal of Contemporary Hospitality Management,* Vol. 5 No. 1, pp. 22-7.

Stayer, R. (1990), "How I learnt to let my workers lead", *Harvard Business Review,* November/December.

Ullmann, W. and Wonnemann, T. (1994), "Swatch's worldwide launch of a database club: a case study", *Journal of Database Marketing,* Vol. 1 No. 4, pp. 307-13.

Williams, D. (1990), "Selling via mail order? Look beyond the list", *Business Marketing Digest,* Vol. 15 No. 3, pp. 65-8.

Wills, G. (1991), "Enabling customers to drive your enterprise", *European Journal of Marketing,* Vol. 25 No. 4, pp. 199-216.

Wills, G. (1992), "Enabling managerial growth and ownership succession", *Management Decision,* Vol. 30 No. 1, pp. 10-26.

Wills, G. and Wills, J. (1992), "Journey to marketing clubland", *Marketing Intelligence & Planning,* Vol. 10 No. 2, pp. 22-36 (see Chapter 12).

Wills, G. and Wills, J. (1993), "Learning in marketing clubland", *Marketing Intelligence & Planning,* Vol. 11 No. 11, pp. 31-46 (see Chapter 13).

Wills, G., Bruce, B. and Duncan, T. (1991), "Creating a marketing intelligentsia", *Marketing Intelligence & Planning,* Vol. 9 No. 4, pp. 1-20 (see Chapter 11).

Wood, A. (1993), "Efficient consumer response", *Logistics Information Management,* Vol. 6 No. 4, pp. 38-40.

Woods, B. (1993), "Hidden assets", *Direct Response* (UK), July, pp. 30-1.

Woolf, B.P. (1993), "Reward the right customer", *International Trends in Retailing,* Vol. 10 No. 1, Summer, pp. 39-40.